T0271386

Pollution Across Borders

Transboundary Fire, Smoke and Haze in Southeast Asia

Pollution Across Borders

Transboundary Fire, Smoke and Haze in Southeast Asia

Editor

Euston Quah
Nanyang Technological University, Singapore

with

Tsiat Siong Tan
Nanyang Technological University, Singapore

NEW JERSEY · LONDON · SINGAPORE · BEIJING · SHANGHAI · HONG KONG · TAIPEI · CHENNAI · TOKYO

Published by

World Scientific Publishing Co. Pte. Ltd.

5 Toh Tuck Link, Singapore 596224

USA office: 27 Warren Street, Suite 401-402, Hackensack, NJ 07601

UK office: 57 Shelton Street, Covent Garden, London WC2H 9HE

Library of Congress Cataloging-in-Publication Data

Names: Quah, Euston, editor.

Title: Pollution across borders : transboundary fire, smoke and haze in Southeast Asia /
   [edited by] Euston Quah.

Description: New Jersey : World Scientific, [2017] |
   Includes bibliographical references and index.

Identifiers: LCCN 2017015026 | ISBN 9789813203914 (hardcover)

Subjects: LCSH: Transboundary pollution--Southeast Asia. | Air--Pollution--Southeast Asia. |
   Smaze--Southeast Asia. | Haze--Southeast Asia. |
   Burning of land--Environmental aspects--Southeast Asia.

Classification: LCC TD883.7.A78 P65 2017 | DDC 363.738/70959--dc23

LC record available at https://lccn.loc.gov/2017015026

British Library Cataloguing-in-Publication Data

A catalogue record for this book is available from the British Library.

Desk Editor: Jiang Yulin

Typeset by Stallion Press
Email: enquiries@stallionpress.com

Printed in Singapore

About the Editors

Euston Quah

Professor Euston Quah is Head of Economics at the Nanyang Technological University (NTU), Singapore, and an Adjunct Principal Research Fellow of the Institute of Policy Studies at the National University of Singapore (NUS). He was formerly Chair, School of Humanities and Social Sciences at NTU; Vice-Dean, Faculty of Arts and Social Sciences; Deputy Director of the Public Policy Program (now called the Lee Kuan Yew School of Public Policy); and headed the economics department at NUS. A prolific writer, Professor Quah had published over 100 papers in major internationally refereed journals and opinion pieces. His most recent works include a paper in an international publication on cost-benefit analysis for Oxford University Press, 2013; a Lead Journal Article in *World Economy* in 2015; a Commentary in the *Asian Economic Policy Review* in 2016; an Invited Paper in the *Macroeconomic Review* in 2017; and a forthcoming paper in the *Annual Review of Resource Economics*, 2018. Two books on Cost-Benefit Analysis was published by Routledge, UK in 2007 and 2012 respectively. His work on *Cost-Benefit Analysis* (with E.J. Mishan 5th edition, and 6th edition due in 2019) was recommended for reference by the US White House, Office of Management and Budget for use by Government Agencies applying for project grants. He was co-author of an Asian Edition of the best selling *Principles of Economics* text with Gregory Mankiw of Harvard University, now a second edition in 2013. The Third Edition will be published in 2019.

Professor Quah advises the Singapore Government in various Ministries and was a Member of the Prime Minister's Economic Strategies Sub-Committee on Energy and the Environment. He had served on the Boards of Energy Market Authority, Fare Review Mechanism Committee of the Ministry of Transport, Board Member of the Energy Studies Institute at NUS; and presently sits on the Government's Market Compliance and Surveillance Committee; the Singapore Medical Council's Complaints Committee; the Advisory Panel of the Ministry of Finance; and the Competition and Consumer Commission of Singapore. In 2016, Professor Quah was appointed a Member of the Social Sciences Research Council of Singapore. He is also a Review Panel Member for the Bill and Melinda Gates Foundation project hosted by the Overseas Development Institute, London ; and in 2015 was inducted as a Fellow Member of the prestigious learned society, European Academy of Science and Arts. Professor Quah is Editor of the *Singapore Economic Review* (since 2002), and the President of the Economic Society of Singapore since 2009. He has been invited by Stanford University, Princeton University, the USA Inter-Pacific Bar Association, WWF for Asia, UNESCAP, Earth Institute of Columbia University (Asian Meetings), ADBI and ADB to speak at their functions and conferences and he is one of the most highly cited and influential university economists in Singapore.

Tsiat Siong Tan

Tsiat Siong Tan is a researcher with the Future Resilient Systems (FRS) project, at Singapore-ETH Centre and Nanyang Technological University.

Preface and Introduction

The Southeast Asian transboundary haze has resulted in severe environmental, health, political and economic ramifications in the region since the 1970s, and has resurfaced time and again in the past decades. In the 1997, 2013 and 2015 episodes, forest fires and haze have brought adverse impacts in the scale of tens of billions of dollars to the affected countries: Singapore, Malaysia, Brunei, the Philippines, Thailand, Indonesia itself, and even to the rest of the world through carbon emissions and climate change. Largely a result of agricultural slash-and-burn practices on peatland in Indonesia, the pollution problem remains largely intractable and complex despite repeated efforts to mitigate it.

Resolving this cross-border issue is critical and yet more insurmountable than it appears. It fundamentally is a classic problem of public good and common property, where everyone owns the commons, yet no one is compelled to be fully responsible for it. However, standard economic tools such as taxes or a simple Coasian solution cannot be applied. Indonesia can be pressured but not forced to reduce its pollution.

The problem is complex to address, primarily due to its transboundary nature that has made it difficult to assert extra-territorial jurisdiction without infringing on the sovereignty of the culprit nation. For Indonesia, enacting land-use statutes, changing regulatory institutions, and enforcing laws are complex and tedious. Plantation owners often deny using fire to clear land, and blame shifting cultivators for starting fires in their small holdings, which later spread to plantations. Proof of negligence must be shown, which is susceptible to delays and transaction costs. Moreover, related to the enforcement problems is the complicated nature of Indonesia's decentralized governance system. The coordination of responsibility for forest fires and haze is spread unevenly across many central and local agencies, with many overlaps.

For all affected countries, it is primarily important to undertake the valuation of the haze's negative impact costs, such that aid can be offered to Indonesia, and assistance can be provided to and targeted at sectors hurt by the haze. One approach could be to spend a sum not exceeding the costs of the haze to enhance the ability of Indonesian authorities to detect, locate and respond to the fires, as well as strengthen its ability to prosecute those responsible. Another approach could involve victim countries paying subsidies to Indonesian plantation owners and farmers to encourage land clearing by non-burning methods as a form of external or international development aid.

We must continue to require the involvement and coordination of all stakeholders, and to search for interdisciplinary solutions from the fields of economics, business, law, environmental science, engineering, political science, geography and others.

This book has been put together with the aim of providing a holistic set of perspectives to aid the understanding and possible resolution of the Indonesian forest fires and transboundary haze. Characterized by their varied approaches and diverse ideas, more than 20 experts from the region have contributed an extensive range of insights to the subject. The essays range from in-depth and technical deliberations, to a balanced exploration of key issues suitable for classroom discussions and for case studies.

This collection of articles includes the facts and history of the forest fires and haze, challenges in achieving the twin goals of economic development and environmental conservation and preservation, and the evaluation and suggestions of public policies. You will also find perceptive analysis through the multifaceted lens of game theory, free-rider problem, the Polluter-Pay-Principle and the Victim-Pay-Principle, investment decisions, international law and cooperation, the ASEAN Agreement on Transboundary Haze, Singapore's Transboundary Haze Pollution Act, technological innovations, Corporate Social Responsibility and Public-Private Partnerships, industry specific initiatives, the role of government and environmental governance, environmental economic valuation and damage assessments, forestry management, the road to sustainability, mitigation of and adaptation to climate change, political economy , and more, including nature's role in this.

From these articles, the reader will be able to develop a clearer view of the complexity and magnitude of the transboundary haze problem, and realize one possible or feasible approach to this persistent cross-border problem. And with optimism, it is in this combination of proposed measures and strategies that we hope to find the elusive solution we have always been searching for.

Professor Euston Quah and Tsiat Siong Tan

Contents

1 Managing Indonesian Haze: Complexities and Challenges

Asit K. Biswas

Distinguished Visiting Professor, Lee Kuan Yew School of Public Policy,
National University of Singapore
prof.asit.k.biswas@gmail.com

Cecilia Tortajada

Senior Research Fellow, Institute of Water Policy,
Lee Kuan Yew School of Public Policy, National University of Singapore
cecilia.tortajada@nus.edu.sg

Introduction

Transboundary air pollution is not a new phenomenon. It has existed for centuries in one form or another. Causes for this pollution could be both natural and man-made. However, prior to 1960, it was not a serious issue since population levels, urbanization rates and the extent of industrial and agricultural activities were rather low. Equally, prior to 1960, all types of environmental pollution were generally accepted by societies to be prices that had to be paid for economic progress. In addition, the health, economic and environmental costs of air, water and land pollution were neither properly known nor fully appreciated. Thus, if there was a conflict between economic growth and environmental pollution, it was routinely resolved in favor of economic growth.

The situation started to change around 1970 when many people in the industrialized world started to realize and appreciate the increasing costs of environmental pollution that had significantly adverse impacts on the human and the ecosystems health. The smog of the 1950s in major cities like London, Los Angeles and Tokyo made their inhabitants question if unfettered economic growth, at the cost of environmental deterioration, is a desirable long-term societal policy option. Shortly thereafter, a nascent environmental movement started. It began to gain momentum and flexed its muscles during the post-1970 period.

Transboundary air pollution first burst into global consciousness during the 1960s when some Swedish and Norwegian scientists noted that a large number of their lakes were showing increasing acidity. This resulted in a steady increase in fish kills and serious negative impacts on many other species. Aquatic biodiversity was declining. Forests were also adversely affected.

The culprit was found to be acid rain whose sources were outside Scandinavia. Acid rain was being caused by sulfur dioxide and nitrogen oxides emitted by factories in the United Kingdom, West Germany and several East European countries which reacted in the atmosphere with the water droplets in the clouds. The net result was that rainfall in Sweden and Norway contained higher concentrations of nitric and sulfuric acids, which started to take its toll on aquatic and territorial ecosystems by acidifying lake waters.

During the late 1960s, Sweden tried very hard to convince the emitting countries of the problems they were causing in Scandinavia and attempted to persuade them to reduce their atmospheric emissions. Being unable to move bigger countries like the United Kingdom or Germany to reduce their emissions, Sweden offered to host the 1972 United Nations World Conference on the Human Environment in Stockholm. The hidden agenda behind this offer was to bring the problem of transboundary air pollution and the harm it was causing to Sweden and other Scandinavian countries to global attention.

This strategy succeeded in increasing the global awareness of this issue (Biswas and Tortajada, 2013) and also internationalizing and bringing it to a major United Nations forum. This Conference and subsequent efforts led to the adoption of the Convention on Long-range Transboundary Air Pollution in 1979 (UNECE, 1979). It entered into force in 1983. This has now been ratified by 53 members of the UN Economic Commission for Europe.

As a result of this legally binding treaty, collective efforts of the emissions of harmful air pollutants have been reduced by 40 to 80 percent since 1990. Sulphur emissions have declined very significantly. This meant that acid rain problems in Europe had been mostly consigned to the dustbin of history, and forests are healthier.

Some 30 years later, transboundary smoke pollution became an important concern in most Southeast Asian countries, except for Myanmar and Laos, and some Pacific island areas like Guam, Palau and Northern Mariana Islands. They are affected by haze because of forest and vegetation clearance by fire, primarily in the Indonesian provinces of Sumatra and Kalimantan. This clearance is to ensure increasingly more and more land is available for expanding the area for palm oil cultivation to meet the burgeoning global demand.

The main difference between the Swedish and the Indonesian cases is that acid rain had very limited impact on the emitting countries like United Kingdom and Germany. In contrast, the Indonesian haze is having the most severe health, economic and environmental impact on the country from where it originates, as well as on its neighbors. Even then, it has proved to be an intractable problem to solve.

Indonesian haze: Background to the problem

Palm oil is not indigenous to Asia. The Portuguese first discovered palm oil in the 15th century in Africa. They found that the small farmers of West African rainforests used palm oil for soups and baking. Following this Portuguese discovery, palm oil gradually became an important provision for trading caravans and slave ships.

In an Asian context, the British administrators first introduced oil palm in the 1830s to India. They planted some seedlings in the botanic garden of Kolkata. Trial commercial planting was started later in Kerala. They did not make much headway.

In Indonesia, oil palm first arrived in 1848, via Amsterdam. Like in India, four seedlings were planted in the botanic garden of Bogor. They were initially considered to be ornamental plants and were planted in tobacco estates and roadsides for beautification purposes.

Commercial plantation for palm oil in Indonesia started in 1911, in Sumatra, by a Belgian agronomist, Adrien Hallet. Hallet had interest in rubber plantations in the then Belgian Congo. In 1912, Henri Fauconnier, a Frenchman, bought some seedlings from Hallet. He established the first commercial plantation in Malaysia, at Tennamaram Estate, Selangor. Having failed at making a successful coffee plantation, Fauconnier thought he would have better luck with palm oil.

After the Second World War, rubber plantations ran into severe headwinds because of rapidly falling demands. Large-scale plantation owners looked at crop diversification, and oil palm appeared to be a good potential alternative. An intensive agricultural research and breeding program was started in Malaysia to assess the economic feasibility of palm oil.

Because of good agro-climatic conditions, oil palm cultivation has always been a good prospect for Malaysia and Indonesia. However, it has had a rather checkered development history in Indonesia. Its development was in the doldrums after Indonesia's independence in 1945. Dutch plantation owners no longer received financial and political support from the colonial government. Some two decades later, in 1967, commercial plantations picked up some steam under its second President, Suharto. Investments were made through state-owned enterprises, with financial support from the World Bank.

During the 1980s, oil palm plantations received a major boost following a finding that these trees are pollinated by a tiny weevil, and not by wind, as previously thought. Introduction of these pollinating weevils in Malaysia and Indonesia dramatically reduced the cost of pollination which earlier was being done by human hands. In the 1980s, Malaysia became the world's largest producer of palm oil.

With an increasing global population and more and more people having steadily higher incomes, their dietary habits and requirements have changed as well. The global demand for edible oils and fats have steadily increased following the Second World War. With this ever increasing demand, global production of all types of oil seed crops have increased concomitantly as well.

Oil palm is the most efficient oilseed crop in the world. Its average oil yield per hectare is about 6–10 times that of soybean, sunflower or rapeseed (Figure 1). At present the most efficient oilseed farmers can get as much as 8 metric tons of oil per hectare. Thus, not surprisingly, even though oil palm accounts for less than 6% of total area in which ten of the world's most important major oil seed crops are cultivated, it accounts for nearly 32% of global oil and fat outputs.

Currently palm oil is Indonesia's second most successful agricultural product, after paddy rice. It is now by far the country's largest agricultural export earner, accounting for nearly 11% of the country's total export

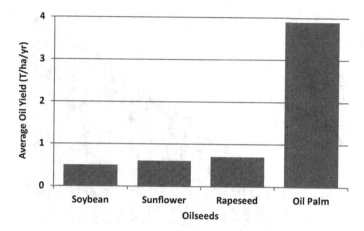

Figure 1.   Average Yield of Oilseeds Crops
Source: Oil World (2013).

earnings. For the rural poor, it is a most important source of income and survival. In 2015, 54% of the Indonesians lived in rural areas. Thus, oil palm cultivation has been one of the important pillars for economic development and poverty alleviation policy for the country in the recent decades.

Development of oil palm cultivation truly picked up steam when Susilo Bambang Yudhoyono was re-elected as the Indonesian President. In his inauguration speech in 2009, he expressed his determination to improve the country's welfare and reduce poverty. These objectives, he said, would be achieved through "economic development based on competitiveness of natural resources management and human resources development" (Jakarta Globe, http://jakartaglobe.beritasatu.com/archive/sbys-inaugural-speech-the-text/).

President Yudhoyono's plan led Indonesia to become the largest palm oil producer country by far in the world. In 2016, its annual production was estimated at 35 million metric tons (MT). The second largest producer, Malaysia, produces 21 million MT, and the third, Thailand, produces 2.3 million MT, only a small fraction of Indonesia. The phenomenal increase in production meant that the land area under oil palm cultivation during 1995–2016 expanded very significantly as well. This steady increase in area is shown in Figure 2.

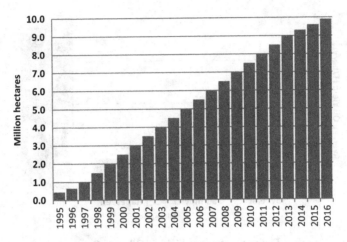

Figure 2.   Indonesian Oil Palm Cultivated Areas, 1995–2016

Source: USDA (2015).

Under Yudhoyono's Presidency, palm oil production became one of the main pillars of economic development and poverty alleviation, especially for the rural areas. In 2008, just before he was elected President, smallholders accounted for nearly 40% of the land, some 1.2 million ha, where oil palm was cultivated. Some 3.5 million households, or about 15 million rural Indonesians, made a living out of growing oil palm. The number of people involved in palm oil supply chain activities (off-farm) was even greater. Thus, palm oil was already an important economic activity in the rural areas before he started his second term as the President.

The plan for agricultural expansion of oil palm was truly ambitious. It proposed to increase the cultivated area to 10–12 million ha by 2020. This would produce 40 million MT of crude palm oil. This expansion was expected to generate new employment opportunities for 1.3 million households. If so, it would reduce the total number of poor people by at least 5.2 million, from an estimated 30 million in 2009.

Favored by government policies, economics of production, commercial logic, and accelerating global demands, Indonesia overtook Malaysia as the world's largest commercial palm oil producer. The increase in palm oil production in Indonesia, during the 1964 to 2016 period, has been truly phenomenal. In 1964, Indonesia produced only 157,000 MT of palm oil.

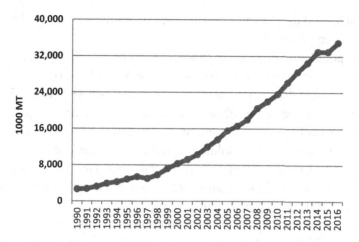

Figure 3.   Indonesian Palm Oil Production

Source: Index Mundi (2016).

It increased to 752,000 MT in 1980, to 2.65 million MT by 1990, 8.3 million MT in 2000, 23.6 million MT in 2010 and 35.0 million MT by 2016. The growth in production is shown in Figure 3.

Not surprisingly this phenomenal growth in palm oil cultivation during the post-1990 period (Figure 3) has been possible due to an exponential increase in land areas where oil palm had been planted (Figure 2). During the 1995 to 2016 period alone, the area planted increased by some 20 times, and quantity produced by some eleven times. Viewed from any perspective, these growth statistics have been phenomenal.

Implications of this growth

The exponential increase in land area used for oil palm cultivation in Indonesia has been possible primarily through the clearance of primary and secondary forests and other vegetative growths. This of course is not new. As the population of Indonesia has increased, there has been progressive deforestation, so that enough land could be made available for all kinds of human activities, including agriculture. In Indonesia's case, the pressure on land clearance has been intense and sustained, especially during the last 50 years, because of oil palm, rubber and timber plantations. The problems have been further exacerbated by the absence of any reasonable land use planning and absence of good environmental policies and management practices.

The population pressure in Indonesia can be visualized from the following facts. In 1960, its population was 88.69 million. By 2015, it had increased to 255.46 million. In other words, in a period of only 55 years, its population increased by about three times. In 1960, 85% of its population (75.38 million) lived in rural areas. By 2015, while this percentage had declined to 46%, in absolute number its rural population was much higher (117.51 million) than what it was 50 years ago. An overwhelming percentage of the rural population still depends on agriculture to earn a living, and still live below the poverty line.

Spurred by steadily increasing global demands for palm oil, timber and rubber, Indonesia has progressively lost forest cover to agricultural development due to deliberate and intentional burning (Quah, 2002). Much of this forest loss has taken place in the lowlands of Sumatra and Kalimantan (Mayer, 2006). This region witnessed an annual deforestation rate of nearly 3.5% during the 1990s. Hansen *et al.* (2009) have estimated that between 1990 and 2005, this region lost nearly 40% of its forest cover. Margono *et al.* (2014) have estimated that more than 6Mha of deforestation took place in Indonesia between 2001 and 2012. By 2012, deforestation in Indonesia exceeded that of Brazil.

Several recent analyses have found that the latest trend has been to move away from the clearance of dryland forests to secondary and peatland forests (Margono *et al.*, 2014; Miettinen *et al.*, 2011). Fire is used extensively to clear peatland and degraded forests. Herein lies the genesis of a major transboundary air pollution problem because of economic necessity of cultivating oil palm in Indonesia.

Peatland forests are generally protected by the presence of high water tables. In order to reclaim these lands for planting oil palm trees, water has to be drained so that vegetation growth can be cleared by fire. As the water table declines, forests and vegetation become increasingly susceptible to burning, as well as to oxidation. This burning has now become the primary source of national and transboundary air pollution. Oxidation does not have any perceptible impact on air quality.

Peatlands are major carbon sinks and can store up to 20 times more carbon dioxide compared to tropical rainforests on normal mineral soils.

Thus, when they are burnt, they become major emitters of CO_2 as well as other undesirable gaseous pollutants like CO, CH_4, NO_x and NH_3. Minister Zulkifli (2015) of Singapore noted during his speech at UNFCCC COP-21 meeting in Paris that peat fires, in 2015, in Indonesia, released over one giga-ton of CO_2 into the atmosphere. These emissions, only over a few months, were higher than the total annual carbon emissions of a major industrialized country like Germany.

This burning, in addition to producing greenhouse gases, contributes to extensive generation of smoke, including particulate pollution, which in turn contributes to serious haze problems in Indonesia and neighboring countries every summer.

The direct linkages between the demand for plantation products, global economy and deforestation can be appreciated by the fact that deforestation rates were significantly lower during the economically lean years of 1999–2001. During this period, the demand for commodities like palm oil was not robust. As the global economy improved, demand increased, and prices of palm oil and other commodities firmed as well. This, in turn, increased the production of palm oil in Indonesia.

Traditionally, forests and degraded lands are mostly cleared by slash-and-burn agriculture. This has been the normal practice in the country for centuries. The process is cheap, easy and does not require any technical skill or heavy machineries. The fire also helps to control pest. The ash and decomposition of organic matters improve soil fertility, at least over the short-term. Because of low costs, convenience and some perceived benefits, fire is widely used as an effective means to clear vegetative growth.

Palm oil production and haze

The forest clearance for agricultural activities takes place every year during the dry season between July and September. The smoke created by these forest fires not only engulfs the provinces where they originate, but also other Indonesian provinces and neighboring countries. During droughts and El Niño years, when the biomass is very dry and rainfall is scarce, the haze problems in Indonesia and neighboring countries become significantly worse. It should be noted that serious drought years are not necessarily El Niño years.

During the 1970s and 1980s, the haze issue was neither serious nor extensive. In addition, the health, economic and environmental impacts were neither well-known nor serious enough so that the public and the policymakers were concerned (Koplitz *et al.*, 2016). By 1995, however, the haze problem was deemed to have become a serious environmental and health issues over a large region, especially as it recurred each summer. This forced the ASEAN environment ministers to take up the issue of transboundary haze problems.

The particulate and gaseous emissions from extensive and intensive forest, vegetation and peat fires of 1997 were especially severe. It was an El Niño year when severe drought and dry vegetation conditions ensured that the land-clearing fires, once started, became uncontrollable. Liew *et al.* (1998) estimated the burned areas in Kalimantan and Sumatra based on SPOT images. This estimate showed that of actual burned area, 50% came from agriculture and plantation, 30% from forests and bushes and 20% from peat forests. These fires were an important source of particulate and gaseous emissions to the local, national and regional atmosphere. In fact, these fires contributed significantly to more particulate and gaseous emissions to the atmosphere compared to even the unprecedented Kuwaiti oil fire of 1991 (Levine, 1999) which was a major carbon emitter.

The continuation of haze every summer and the serious health and economic impacts of the 1997 haze galvanized all the neighboring countries of Indonesia into action. They were not only getting exasperated by the regular annual occurrence of haze but were also becoming increasingly aware of the health risks, initially of PM_{10} and a decade later of $PM_{2.5}$. The media and social pressure on the governments of these countries to do something significant to reduce the extent and magnitude of the haze that originated from Indonesia became increasingly intense.

The neighboring governments took up the haze issue bilaterally with the Indonesian Government as well as multilaterally through the ASEAN. These discussions led to an Agreement on Transboundary Haze Pollution which was signed in Kuala Lumpur, on 10 June 2002 (http://haze.asean.org/asean-agreement-on-transboundary-haze-pollution/). The Agreement was ratified by sufficient number of countries to enter into force on 25 November 2003.

Even though the Agreement entered into force in late 2003, Indonesia was not a signatory to it for over a decade due to various political maneuvering (Biswas and Hartley, 2015). The Indonesian Parliament refused to endorse the Agreement. It was finally ratified more than a decade later, on 14 October 2014.

Although the ASEAN Agreement has been in force since 2003, and Indonesia finally ratified it in 2014, it does not appear to have had any perceptible effects in reducing the impacts of the 2015 haze. This was another bad year for serious transboundary haze. There are many reasons for this rather sad state of affairs.

The first set of reasons relate to the ASEAN haze treaty and its possible effectiveness in reducing haze. There is no question that all the ASEAN countries appreciate the adverse health, economic and environmental impacts of the transboundary haze. Equally, it is now firmly in their agenda as a priority issue to be tackled. While the ASEAN rhetoric has been strong, real actions on the ground thus far have been minimal. In the consensus-driven world of the ASEAN, with a strong history of non-intervention in the affairs of member countries, finding an effective and implementable solution for a transboundary issue like the haze has caught the Association flat-footed.

The problem has been exacerbated by the fact that under the ASEAN protocol, any assistance to a country can only be by unanimous consent and this must have the agreement of the recipient state. Indonesia has consistently denied most offers of assistance, probably because of sovereignty reasons as well as political and reputational implications within and outside the country and national pride. Furthermore, the Agreement may be legally binding but it has no enforceability, accountability, or any mechanism for independent monitoring, sanctions on the offending countries, or even dispute settlement procedures. The Agreement can only be effective if all the 10 countries concurrently and voluntarily take all the necessary measures to reduce haze. This is somewhat unlikely and often an unrealistic expectation.

The second set of reasons has to do with Indonesia directly. The country faces serious challenges and constraints to reducing haze. Among these are the following:

- *Magnitude and extent of the problem* — Given the vast area of the country, its archipelagic nature, sheer size and extent of forest areas, number of states involved from where fire originates, and immense number of parties involved in palm oil cultivation, any objective and realistic analysis will indicate that the problem cannot be solved over the short-term. The current situation has developed over decades, promoted by proactive palm oil production policies and strategies of the government and international institutions. These policies never seriously considered the potential environment health and social impacts of increasing palm oil production. It will be years before the situation can be brought fully under control.

- *Centre–state–municipal relations* — The ASEAN Agreement was eventually ratified by the Central Government of Indonesia, but land use, forest management, agriculture, etc., are state subjects on which the center has at best limited say or control. In addition, some of the issues are under the jurisdiction of the municipal governments over which neither the central nor state government institutions have much control.

- *Appropriate policies of different institutions at three government levels* — Effective haze management requires coherent policies at all the three governmental levels. Such cohesion is simply absent at present. Even at the Federal level, good collaboration and coordination will be necessary between departments and ministries like land, water, environment, agriculture, industry, commerce and law, all of which have a say in certain aspects of haze management. Even all developed countries, let alone a developing country like Indonesia, have found it very difficult to have such a coordinated policy approach on a multi-sectoral and multi-issue problem. To get all the institutions, at all the three levels, in many different states, marching towards the same policy goals, at reasonably similar speeds, will be a herculean, if not an impossible, task under the best of circumstances.

- *Institutional capacities* — Adequate technical, managerial, administrative, financial, legal and regulatory capacities (Atkinson, 2014) at all the three levels leave much to be desired. It will take some time before such effective capacities can be developed.

- *Corruption* — This is endemic at all the three levels of the government. It will be extremely difficult to reduce it significantly in the foreseeable future so that significant reductions in haze can be observed.

- *Powerful interests* — While haze is a serious problem in Indonesia and its neighboring countries, for farmers and industries, burning continues to be the simplest and the cheapest method to clear the vegetation. Weaning them from such practices will require the development of newer and economic ways to clear forest and vegetation growth, monitoring of prevalent situations, enforcement of laws and regulations and massive education and public awareness programs on the severe economic, environmental, health and social impacts of haze, both within and outside the region from where fire originates. All of these are mostly missing at present.
- *Pervasive poverty* — For millions of poor and small-scale rural farmers, burning has been the traditional way to clear land to produce palm oil. Their livelihoods depend on this practice. For any democracy, where candidates depend on the votes of rural farmers to get elected, it will be a very difficult political choice to ask this group to use an alternative methods of vegetation clearance which are more expensive and complex. The issue of how to provide the rural farmers the means to earn sustainable livelihoods and alleviate their levels of poverty have been mostly missed by Western scholars and development experts. Effective and economic ways have to be found for these poor rural farmers if the incidence of haze has to be substantially reduced.

Given all the constraints and challenges faced by Indonesia, it is highly unlikely that the haze situation can be significantly reduced before 2025 at the earliest. The roadmap that the ASEAN countries adopted in Kuala Lumpur, in August 2016, to achieve a haze-free ASEAN zone by 2020 will most likely prove to be a pipe-dream.

Indonesia is the main emitter of the smoke. Thus, most of the important decisions to control smoke have to be taken by the Indonesian Government at all the three levels. These policies then have to be implemented within a reasonable period of time so that the ASEAN region could become a haze-free zone over the long term. This objective, laudable though it is, is neither realistic nor achievable. In fact, even at the time of writing this contribution in August 2016, a realistic set of implementable policies that can substantially reduce smoke emissions from Indonesia by 2020 does not even exist, let alone consideration of their potential implementation.

Tellingly, the Indonesian Minister for Environment and Natural Resources, Siti Nurbaya Bakar, was not even present at the ASEAN meeting in Kuala Lumpur in August 2016, ostensibly because she was occupied with a conservation meeting in Bali. This was the second year in a row that the Minister had missed the meeting of the ASEAN Environment Ministers. Under such political conditions, it is difficult to be hopeful. It does not convey the right message in terms of commitment, given her earlier combative responses to Singapore's concerns on transboundary haze. For example, in October 2015, she said: "I would respectfully ask them (Singaporean neighbors) to stop making so many comments, particularly when it comes to the fires and haze-related issues." (Must Share News, http://mustsharenews.com/indonesian-minister-haze/).

In addition, in early 2014, the Indonesia Vice-President, Jusuf Kalla, said, "For 11 months, they (our neighbors) enjoyed nice air from Indonesia and they never thanked us. They have suffered because of the haze for one month and they get upset" (Jakarta Globe, http://jakartaglobe.beritasatu.com/news/vp-kalla-slams-neighboring-countries-over-haze-complaints/). He also said that the haze that affected Malaysia and Singapore was blown there by wind, over which Indonesia has no control. Thus, the mindsets at high political levels need to substantially change for effective haze management.

It is thus difficult to be optimistic that the transboundary haze can be controlled reasonably well by 2020, let alone transforming the region into a haze-free zone.

Way ahead

There is no question that over the last five decades the problem of transboundary air pollution due to the Southeast Asian haze has become progressively more serious. During the 1960s and the 1970s, the problem did exist. However, it was minor and thus did not receive much attention in the countries that were affected by it. It was neither a political nor a social issue, and did not attract any serious political, public or media attention.

The situation started to change in the late 1980s, and by the first half of the 1990s, transboundary haze problems could no longer be ignored. As the area under palm oil dramatically increased during the 1990s and the post-2000 periods, it became an important public policy issue in Malaysia and Singapore. The environmental, economic and health impacts of the transboundary haze are now very serious, and have also become important political, public and media concerns. These cannot longer be ignored.

Indonesia is the country which suffers the most from the haze. It is also the country which benefits the most because of the palm oil production. The benefits it brings to the country in terms of export earning, employment generation and providing livelihoods for millions of its households are substantial. In contrast, countries like Singapore and Malaysia receive no benefits but pay heavy costs due to a haze that originates in another country on which they have only a limited say and virtually no control.

Because of the lopsided distribution of benefits and costs, views and perspectives of Indonesia and the haze-affected countries have often been different (Biswas & Tortajada, 2016). Harsh words have been exchanged in recent years between the politicians of the concerned countries. This antagonism and frustration, though understandable, is unlikely to improve the situation.

Any lasting and long-term solution will have to start with the fact that Indonesia will not only have to continue with the production of palm oil in the future but also may have to increase it substantially. Politically and economically, it has few other viable alternatives to maintain and generate additional employment in rural areas and palm oil supply chain activities for its steadily increasing population.

Future solutions involving expanding palm oil production, while significantly reducing current haze levels in Indonesia and its neighboring countries, will lie on where the expansion takes place and how well the production processes are managed in terms of their environmental footprints, especially in terms of generation of smoke, based on the best available knowledge and information. For example, estimates by Marlier *et al.* (2015) indicate that if the peatland in Sumatra and Kalimantan can be properly protected, it could reduce future emissions by some 60% compared to current practices. This could reduce smoke concentration levels by a factor of three in Singapore.

Similarly, if the current moratorium on new concessions in peatland and primary forests can be maintained, it could reduce emissions by 10%.

Another 35% reduction will be possible by limiting expansions on all peat and forest lands (Austin *et al.*, 2015).

Thus, solutions do exist which could enable Indonesia to expand its palm oil production significantly up to 2050, and concurrently reduce smoke generation very substantially. This would require that future expansions be carefully planned and managed which would enable the country to double its palm oil production while reducing the existing levels of transboundary haze. This two-prong policy strategy is most certainly achievable and will benefit both Indonesia and its neighbors.

In order to make progress in solving this complex and recurring annual challenge, it is imperative that the Indonesian public and policy-makers be convinced beyond doubt that haze is bad for them. This will require credible and objective evidence. Even though the haze has been around for decades, no reliable estimates are currently available as to the costs to Indonesia in terms of health, loss of productivity or impacts on foreign direct investments. Foreign investors may decide to avoid the country for concerns not only with macroeconomic conditions, inadequate infrastructure, effectiveness of legal systems and traffic congestions but also because of extensive air and water pollution. A young and expanding population means the country will have to generate hundreds of thousands of new and well-paid jobs each year. This will require sustained FDI inflows. This effort will be seriously compromised unless environmental and economic conditions are substantially improved. External political pressures alone to address such concerns will be limited without greater public awareness of the problems in Indonesia (Biswas and Hartley, 2015).

The policy battlefield needs to move from the banquet tables of regional summits to the hearts and minds of the Indonesians. Winning their support for stricter regulations and enforcement is a bipartite strategy. First, an effort must be made to gather data and generate robust evidence about the domestic health, economic and environmental impacts of Indonesian haze. Second, the findings must be expressed in an easily understandable way and disseminated widely through the media and other conduits. This should not be a one-off publicity effort, but a sustained awareness initiative that targets the general public and enlists the support of advocacy groups, NGOs, and ultimately (through domestic political pressure) government officials (Herawati and Santoso, 2011).

Public understanding of the haze crisis in Indonesia should be as common as basic literacy, and at least as embedded as the public's knowledge of common health and safety issues related to smoking, pesticides and sanitation (Biswas and Hartley, 2015). This may not produce the instant solution that many believe would result from aggressive diplomatic actions like boycotts and sanctions. Haze mitigation interventions would be far more durable across political cycles because increased awareness often leads to deeply rooted interest in a policy issue. Indeed, an educated Indonesian public may prove to be the most powerful force for change needed to manage the haze effectively in the coming decades.

References

Atkinson, C. L. (2014). Deforestation and transboundary haze in Indonesia: Path dependence and elite influences, *Environment and Urbanization Asia*, 5(2): 253–267.

Austin, K. G., Kasibhatla, P. S., Urban, D. L., Stolle, F. and Vincent, J. (2015). Reconciling oil palm expansion and climate change mitigation in Kalimantan, Indonesia, *PLoS ONE*, 10(5): e0127963. DOI: 10.1371/journal.pone.0127963

Biswas, A. K. and Hartley, K. (2015, September 21). Singapore haze: A new strategy needed. The Diplomat. Retrieved from http://thediplomat.com/2015/09/singapore-haze-a-new-strategy-needed/

Biswas, A. K. and Hartley, K. (2015, September 21). To tackle haze, win over the Indonesian public. *Straits Times*, Singapore.

Biswas, A. K. and Tortajada, C. (2013, August 25). Tackling haze: Learn from the Swedes. *Straits Times*, Singapore.

Biswas, A. K. and Tortajada, C. (2016, August 25). Raise public understanding in Indonesia to tackle haze. *Business Times*, Singapore.

Hansen, M. C., Stehman, S. V., Potapov, P. V., Arunarwati, B., Stolle, F. and Pittman, K. (2009). Quantifying changes in the rates of forest clearing in Indonesia from 1990 to 2005 using remotely sensed data sets, *Environmental Research Letters*, 4(3): 1–12.

Herawati, H. and Santoso, H. (2011). Tropical forest susceptibility to and risk of fire under changing climate: A review of fire nature, policy and institutions in Indonesia, *Forest Policy and Economics*, 13(4): 227–233.

Koplitz, S. N., Mickley, L. J., Marlier, M. E., Buonocore, J. J., Kim, P. S., Liu, T., Sulprizio, M. P., DeFries, R. S., Jacob, D. J., Schwartz, J. and Myers, S. S.

(2016). Public health impacts of the severe haze in equatorial Asia in September — October 2015: A new tool for fire management to reduce downwind smoke exposure in the future, *Environmental Research Letters* (in review).

Levine, J. S. (1999). The 1997 fires in Kalimantan and Sumatra, Indonesia: Gaseous and particulate emissions, *Geophysical Research Letters*, 26(7): 815–818.

Margono, B. A., Potapov, P. V., Turubanova, S., Stolle, F. and Hansen, M. C. (2014). Primary forest cover loss in Indonesia over 2000–2012, *Nature Climate Change*, 4: 730–735.

Marlier, M. E., DeFries, R. S., Kim, P. S., Koplitz, S. N., Jacob, D. J., Mickley, L. J. and Myers, S. S. (2015). Fire emissions and regional air quality impacts from fires in oil palm, timber, and logging concessions in Indonesia, *Environmental Research Letters*, 10(8), 085005.

Mayer, J. (2006). Transboundary perspectives on managing Indonesia's fires, *Journal of Environment and Development*, 15(2): 202–223.

Miettinen, J., Shi, C. and Liew, S. C. (2011). Deforestation rates in insular Southeast Asia between 2000 and 2010, *Global Change Biology*, 17(7): 2261–2270.

Quah, E. (2002). Transboundary pollution in Southeast Asia: The Indonesian fires, *World Development*, 30(3): 429–441.

Zulkifli, M. (2015, December 7). National statement of Singapore to UNFCCC COP-21 meeting, Paris. Ministry for the Environment and Water Resources, Singapore.

2 Transboundary Haze in Southeast Asia: Dealing with Elusive Regional Solutions and Implications on ASEAN Community

Mely Caballero-Anthony

Head, Centre for Non-Traditional Security Studies
Associate Professor, S. Rajaratnam School of International Studies,
Nanyang Technological University
ismcanthony@ntu.edu.sg

ASEAN and its vision of a Community

Since its founding in 1967, the Association of Southeast Asian Nations (ASEAN) has envisioned a regional community that is peaceful, stable, prosperous and harmonious. More than 50 years since its establishment, the idea of how this regional community is envisioned is best articulated in its notion of an ASEAN Community founded on the three pillars: a political and security community, an economic community, and a socio-cultural community. The first pillar, the ASEAN Political-Security Community (APSC) aims to "ensure that countries in the region live at peace with one another and with the world in a just, democratic and harmonious environment." It achieves this by deepening the nature of political cooperation among the 10 ASEAN member states and raising its security cooperation to a higher plane. Mindful that the nature of security challenges facing the region is no longer limited to traditional military threats but is also now increasingly transnational in nature, the APSC puts great emphasis on members states working together in addressing cross-border challenges.

The second pillar, the ASEAN Economic Community (AEC) focuses on regional economic integration characterized by "(a) a single market and production base, (b) a highly competitive economic region, (c) a region of equitable economic development, and (d) a region fully integrated into the global economy." The third pillar, ASEAN Socio-Cultural Community (ASCC) focuses on regional socio-cultural cooperation with the goal of forging a "common identity" and building a "caring and sharing society which is inclusive, and where the well-being, livelihood, and welfare of the peoples are enhanced."[1]

The three components of the ASEAN Community are conceived as being "closely intertwined and mutually reinforcing." According to the *ASEAN Vision 2020*, the creation of an ASEAN Community would ensure closer, more comprehensive and mutually beneficial integration among ASEAN member states and their peoples. It would strengthen the basis for peace, stability, security, development and prosperity within the region in order to enhance regional cooperation against the fast-changing dynamics of the regional and global environment.

While much has been written about the APSC and the AEC, it is only more recently that more attention is now given to the realization of an ASCC. There are a number of reasons why the ASCC could arguably be the most important element in the process of ASEAN Community building and certainly worthy of greater focus.

Firstly, achieving the end goal of the ASEAN Community will be challenging given the region's cultural diversity, different levels of economic development as well as differences in political order and strategic concerns. Managing such diversities in order to achieve equitable development across the region will not be an easy task.

Secondly, the ASCC is fundamental to ASEAN's holistic approach to ASEAN Community building as the goals under this pillar speak to the everyday issues and challenges faced by communities in the region, regardless of socio-economic and political background. One of its imperatives is ASEAN's social agenda that is focused on poverty eradication and human development acknowledging that: (1) "social inequities can threaten economic development and in turn undermine political regimes";

---

[1] See *ASEAN Bali Concord II* and the ASEAN Community Blueprint, (ASEAN Secretariat, 2009).

(2) "economic instability can exacerbate poverty, unemployment, hunger, illness and disease"; and (3) "social instability can emerge from environmental scarcity or the inequitable distribution among stakeholders of the use of environmental assets."[2] The vision of a caring, sharing and people-centered ASEAN is buttressed by four strategic thrusts: social protection systems, environmental sustainability and sustainable resource management, social governance, and the preservation of the region's cultural heritage and cultural identity.[3]

Lastly, within ASEAN's framework of regional cooperation, security and development are always regarded as two sides of the same coin. One cannot have security without development and development without security. In fact, this approach has always characterized ASEAN's conceptualization of regional security. In ASEAN's parlance, security is defined as more than just defending states from external military threats, but also from other equally important threats to the survival of the state — economic underdevelopment, societal divisions, religious conflicts and other issues that threaten the security of the state. Hence, the notion of comprehensive security became ASEAN's organizing security concept and one that puts a premium on building national and regional resilience to achieve security. A key element in building resilience for security is economic development.[4]

Over many years, ASEAN has indeed achieved a certain level of economic development that has made Southeast Asia a much sought after region for trade and foreign direct investments. Despite having gone through a difficult period during the 1997 Asian financial crisis, most of the ASEAN countries have recovered and have since registered stronger economic growth averaging 5 to 6 percent over the last 10 years. Poverty rates have also significantly declined between 2000 and 2010 from 26 percent to 14.7 percent.[5]

---

[2] *The ASEAN Socio-Cultural Community (ASCC) Plan of Action 2004.*

[3] *2004 Vientiane Action Programme 2004–2010*, adopted by the Heads of Government at the 10th ASEAN Summit, Vientiane, Laos, 29 November 2004.

[4] See Alagappa, M. (1988). Comprehensive security: Interpretations in ASEAN Countries, in Scalapino, R. *et al.* (eds), *Asian Security Issues: Regional and Global*, Institute of East Asian Studies, University of California, Berkeley.

[5] ASEAN Statistics., (2015). 'Statistics to tTrack pProgress: ASEAN integration inched up to 2015'. Retrieved, available at http://www.asean.org/storage/images/2013/resources/statistics/statistical_publication/aseanstats_acpms_snapshot_r.pdf, accessed 10 August 2016.

One notes however that despite this economic progress, the difficult challenge of addressing the vicious cycle of underdevelopment and intractable conflicts that often affects many communities, particularly the marginalized and vulnerable communities, remains. And, despite the peace that has prevailed in the region over the last 51 years, and efforts over the last decade to adopt an ASEAN Charter that spells out the norms of democracy, peace and prosperity — there remains significant pockets of insecure communities within states in ASEAN. Moreover, in spite of the emphases on building a "community of caring and sharing societies" through the ASCC, there are many developmental issues that aggravate the insecurities of communities in ASEAN. Among these are access to basic needs like drinking water, decent shelter, medical care and clean air and safe environment.

Transboundary haze pollution and ASEAN's goals of establishing caring and sharing societies

The ASCC Blueprint outlines the characteristics of a caring and sharing ASEAN to include: (1) human development; (2) social welfare and protection; (3) social justice and rights; (4) ensuring environmental sustainability; (5) building an ASEAN Identity; and (6) narrowing the development gap.[6] Against such lofty vision, the perennial problem of transboundary haze pollution that has plagued ASEAN for the last two decades presents a serious challenges and impediment to the realization of an ASEAN Community. One can argue that unless the haze problem is decisively addressed, ASEAN would continue to be faced with credibility problems given the huge gap between its aspirational goals of achieving a caring society and the realities of the region's complex set of environmental and socio-economic issues.

As pointed out by many experts, the haze problem in ASEAN is caused mainly by severe incidences of land and forest fires, made worst by adverse weather conditions. The first worst outbreak of the haze was in 1997 following uncontrolled forest fires in Indonesia's Kalimantan. The urgency of addressing transboundary haze pollution became more

---

[6] *ASEAN Socio-Cultural Community Blueprint* (Jakarta: ASEAN Secretariat, 2009).

acute when this problem inflicted considerable damage to neighbouring countries — Singapore, Malaysia and Brunei. Earlier studies had shown that at the height of the problem, it was estimated that Malaysia and Singapore suffered losses of about 1 billion US dollars in economic activity, such as tourism, air travel and immediate health cost.[7] During the haze period, air quality was monitored constantly in these three states and when the air pollution index (API) reached alarming levels as in cases of Sarawak in Malaysia and parts of Brunei, some people resorted to leaving their states until the haze abated. During that period, the case of the environmental pollution brought on by unregulated slash-and-burn practices of farmers in Kalimantan presented a clear case of the haze posing an existential threat to the lives and security of people in the affected states.

The 1997 haze crisis and its persistence over the years led ASEAN to seriously examine how transboundary issues that affected the well-being and security of peoples of ASEAN were to be managed collectively by its member states. The crisis led to the crafting of the 1997 Regional Haze Action Plan (RHAP) which devised regional cooperation measures in three areas: prevention, monitoring and mitigation of land and forest fires.[8] And in 2002, ASEAN member states adopted the ASEAN Agreement on Transboundary Haze Pollution. Unlike other ASEAN documents, the Haze Agreement was notable in that it is the first legally binding agreement of ASEAN. The RHAP formed the basis of the Haze Agreement. The six-part agreement essentially outlined specific measures for ASEAN member states to take in order to prevent and monitor transboundary haze pollution. Four of the six parts of the Agreement focus on: Monitoring, Assessment, Prevention and Response; Technical Cooperation and Scientific Research; Institutional Arrangements and Procedures.

Although the Agreement came into force in 2003, it took Indonesia 12 years to ratify the Haze Agreement. The delay in Indonesia's ratification of the Haze Agreement was viewed by observers as indicative of the kind of challenges that ASEAN as a regional body faced in generating

---

[7] Tay, S. (2001). The environment and security in Southeast Asia, in Caballero-Anthony, M. and Hassan, M. J. (eds.), *Beyond the Crisis: Challenges and Opportunities*, Institute of Strategic and International Studies: Malaysia, p. 154.

[8] Haze Action Online. (1997). Regional Haze Action Plan. Retrieved http://haze.asean.org/

collective action and commitment to address a shared regional problem. As noted by one analyst, the time it took Indonesia to finally ratify the Haze Agreement in 2014 reflects the view from within that it is "nothing more than signing a non-enforceable agreement."[9]

The signing of the Haze Agreement did not do much to prevent the haze from occurring again. Two of the more severe episodes of transboundary haze occurred in 2013 and in 2015. During these two incidents, the impact on human security and its consequences were more severe.

The impact of the prolonged transboundary haze in 2015[10]

For the affected countries in the region like Singapore, the 2013 haze episode was worrying since it saw the Pollutant Standards Index (PSI) level in the country hit an all-time high of 401. But what made the haze in 2015 more worrying was that it occurred during the El Niño season which has already caused a prolonged dry spell in the region. The El Niño season therefore exacerbated the impact of the forest fires making them last longer than usual.

The 2015 haze led to more visible threats to the human security of the people in Indonesia, Malaysia and Singapore. The most severe effect was on health security, but there were also other economic and environmental impacts that affected the lives of the people in the region. Indonesia had been the worst hit and the immediate suffering and health impacts had pushed Indonesians to protest against forest fires.

Feeling helpless and frustrated with the lack of response from the government, the people in Pekanbaru, Indonesia took to the streets for the first time to demonstrate against the local authorities. The smog was no longer just seen as an annual problem for Singapore and Malaysia; Indonesians were finally calling on their government to recognize the impact of the haze on their health security.

---

[9] Sembiring, M. (2015). Here comes the haze, yet again: Are new measures working?, RSIS Commentary, No. 191, S. Rajaratnam School of International Studies: Singapore.

[10] This section is largely taken from an earlier commentary: Caballero-Anthony, M. and Tian Goh. (2015). ASEAN's haze shroud: Grave threat to human security, RSIS Commentary, No. 207, S. Rajaratnam School of International Studies: Singapore.

A serious health security threat

The World Health Organization (WHO) had warned about the health risks to long-term exposure to particulate matter in the haze, known for short as PM2.5, which are microscopic particles of less than 2.5 micrometers in size. Inhalation of PM2.5 is considered more severe than other pollutants such as PM10 — which are less than 10 micrometers in size.

At the height of the haze problem in 2015, primary and secondary schools were closed in Malaysia and Singapore given the concern over the level of exposure from the pollution level at that time. On 25 September 2015, the recorded PSI level in Singapore then was above 300. In contrast, the situation in Indonesia had reached alarming proportions. Parts of Indonesian in Kalimantan and Sumatra recorded PSI levels above 1,000, with Central Kalimantan hitting a record of 1,995 on 23 September 2015.

Haze-related medical conditions that appeared at that time included acute upper respiratory tract infection, allergies, worsening of asthma and bronchitis, acute conjunctivitis and eczema. According to *Jakarta Post*, there had been a total of 53,428 reported cases of respiratory infections in South Kalimantan, 34,846 in Pekanbaru, 22,855 in South Sumatra, 21,130 in West Kalimantan and 4,121 in Central Kalimantan.

These conditions exclude the potential long-term effects of the haze such as increase in mortality, respiratory and cardiovascular diseases and morbidity as well as reduction in life expectancy. It has been found that children, who are more vulnerable to the ill effects of haze, may suffer from reduced lung development and develop diseases such as asthma. For the poorer communities in the affected regions of the haze, the burden of healthcare cost became more acute.

Socio-economic impact

The economic impacts of the 2015 haze were also significant. In Indonesia, businesses were affected not only by increased absenteeism but by companies having had to close due to the sheer intensity of the haze. Tourism in the region took a hit and airports in Malaysia and Indonesia had to be closed. At least 20 airports in Indonesia were closed since the start of the haze episode. As some countries issued travel warnings to

Singapore, the city-state could suffer a loss of 0.1 percent to 0.4 percent of its gross domestic product (GDP), or at least 3 billion US dollars.[11]

While the haze crisis in 1997 cost Southeast Asia an estimated 9 billion US dollars, the cost in 2015 were more than doubled. According to the World Bank, the 2015 haze cost the government 16 billion US dollars (22.5 billion Singapore dollars) — more than double the sum spent on rebuilding Aceh after the 2004 tsunami.[12] According to the World Bank Indonesian country director, Rodrigo Chaves, it cost 7 billion US dollars to rebuild Indonesia's westernmost province of Aceh after it was hit by a quake-triggered tsunami, with tens of thousands of lives lost, but "the economic impact of the fires [had] has been immense."

Revisiting ASEAN's approaches to transboundary haze: What more needs to be done?

The haze crisis has reached a point where it is no longer one country's problem. It affects the health and human security of everyone. Haze is now as much a human security as well as a national and regional security threat. It is time to tackle it as one of the most immediate, challenging and perennial human security issues in the region. This requires no less than credible commitments and focused efforts at the national and regional levels.

After Indonesia ratified the Haze Agreement in 2014, the government under President Joko Widodo took significant steps in merging the environment and forestry ministries in order to improve bureaucratic coordination and improving efficiency in responding to environmental and forestry issues including forest fires and the consequent haze pollution that occurs almost annually. The government in Jakarta also launched the much anticipated One Map of National Thematic Geospatial Information which is aimed at standardizing information, creating one reference, one

---

[11] O'Callaghan, J. (2015, June 24). Singapore, Malaysia face economic hit from prolonged smog. *Reuters*. Retrieved http://www.reuters.com/article/us-southeastasia-haze-impact-idUSBRE95N0BS20130624

[12] Haze fires cost Indonesia S$22b, twice tsunami bill: World Bank, *The Straits Times*, Singapore, 16 December 2015.

standard, one database, and one portal in order to provide a better avenue for stronger cooperation and coordination for national action.[13]

During the same period, the Singapore parliament passed the Transboundary Haze Pollution Act 2014. The Act aims to deter companies or entities operating in or outside Singapore from taking part in activities that contribute to transboundary haze affecting Singapore. It allows the Singaporean government to impose fines on companies, both local and foreign, that are deemed responsible for unhealthy levels of haze pollution in Singapore. This act was invoked in September 2015, when the Singaporean government launched legal action against five Indonesian companies[14] that involved in setting forest fires in Indonesia that caused haze of hazardous levels in Indonesia, Singapore and Malaysia.[15] This move could lead to massive fines against these companies. It was followed by the issuance of notices to six companies that requested the provision of information on their plans to extinguish and prevent fires on their land. If the companies failed to reply to the notice within the given date, the Singaporean government can arrest representatives of the companies upon the next entry into Singapore.[16]

While a significant development, critics have pointed out that the effectiveness of the new law would largely depend on the ability to accurately identify errant companies or entities operating in Indonesia. Questions about the law enforcement capacity in the Indonesian provinces concerned have also been raised and the cooperation of authorities to go after the culprits. Indonesia's overlapping concession maps have often been cited as one of the main reasons that hinder law enforcement efforts in affected areas. The Indonesian government itself has previously filed suit

---

[13] Sembiring, M. (2015). Here comes the haze, yet again: Are new measures working?, RSIS Commentary, No. 191, S. Rajaratnam School of International Studies: Singapore.

[14] The companies include Asia Pulp and Paper, Rimba Hutani Mas, Sebangun Bumi Andalas Wood Industries, Bumi Sriwijaya Sentosa, and Wachyuni Mandira.

[15] Singapore moves against Indonesian firms over haze, *The Jakarta Post*, 26 September 2015. Retrieved http://www.thejakartapost.com/news/2015/09/26/singapore-moves-against-indonesian-firms-over-haze.html, accessed 15 August 2016.

[16] Hussain, Z. (2016, April 21). Singapore taking action against companies responsible for haze, says Masagos Zulkifli. *The Straits Times*. Retrieved http://www.straitstimes.com/singapore/environment/singapore-taking-action-against-companies-responsible-for-haze-says-masagos, accessed 15 August 2016.

cases against Indonesian companies, but the rulings and implementation have been complicated by aforementioned issues.[17]

As far as the regional Haze Agreement is concerned, more can certainly be done to realize many of the issues covered in the Agreement. Some of these include the following:

- Establish and operationalize the ASEAN Co-ordinating Centre for Transboundary Haze Pollution Control

In the 11th Meeting of the Conference of the Parties to the ASEAN Agreement on Transboundary Haze Pollution on 29 October 2015 in Hanoi, Indonesia's intention of hosting the ASEAN Co-ordinating Centre for Transboundary Haze Pollution Control ("the Centre") was endorsed. ASEAN needs to ensure that adequate resources such as budget, human capital, among others, are in place to support and expedite the establishment of the Centre. While waiting for the Centre to get fully functional, the ASEAN Co-ordinating Centre for Humanitarian Assistance (AHA Centre) has been suggested to take the coordinating role for haze-related joint responses. This temporary arrangement may work in the short run, but considering the complexity of haze issues, a dedicated and fully-operational Centre will significantly help ASEAN to focus on responding more effectively to forest and land fires and addressing the much deeper underlying causes.

- Explore the establishment of a regional mechanism to facilitate law enforcement across ASEAN member states

Problems in enforcing Singapore's Transboundary Haze Pollution Act have been evident recently. While having such a regulation creates room for cross-border legal actions, its implementation seems to be marred by suspicions and distrust between countries. In this regard, ASEAN should capitalize on providing the platform where deliberations on this sensitive matter may take place. A mechanism which enables law enforcers of each ASEAN Member States to cooperate and implement their national laws needs to be agreed.

---

[17]Tan, A. K.-J. (2015). The 'haze' crisis in Southeast Asia: Assessing Singapore's Transboundary Haze Pollution Act, NUS Law Working Paper Series 2015/002, February, http://law.nus.edu.sg/wps/pdfs/002_2015_Alan%20Khee-Jin%20Tan.pdf

- Engage more private sector involvement in managing haze issues

Thus far, regional cooperation on managing the haze problem has involved largely ASEAN officials, working on joint responses, peatland management, and hotspot/haze monitoring. ASEAN has yet to come up with specific guidelines on how it deals with the private sector. A large body of literature has indicated that the private sector plays a critical role in the haze issues. In this regard, ASEAN could meaningfully work with the private sector to assist in solving the issues if it could create guidelines in the following practical issues:

(a) Criteria to issue permits for plantation companies → how many fire-fighters, firefighting equipment and technology, firefighting methods, among others, in relation to the total plantation area that a company needs to possess to ensure sufficient firefighting capability.

(b) Audit system for firefighting efforts → which government agencies are responsible to perform audit, and how many times a year such audit needs to be conducted.

(c) Assistance mechanism for companies not able to meet the minimum standard for firefighting capability → providing loans, rentals, etc., needs to be established.

(d) Exchange of best practices among plantation companies → ASEAN can call for regular meetings and exchanges of best practices among plantation companies which would raise awareness of gold standard and enable norm creation. Additionally, ASEAN needs to think of assistance mechanism that will encourage and enable less financially-abled companies to reach the gold standard.

- Coordinate preventive measures with the finance/banking sectors

Much has been said about dis-incentivizing companies from slash-and-burn practices by curbing their access to capitals obtained from the financial/banking sector. While some tightening has been taking place, the effects of such measures at the regional level remain unclear. Different approaches in different countries may result in less-than-effective outcomes. At regional level, ASEAN could work on establishing harmonized guidelines for the financial actors in the region to come together and act against errant companies in similar manner.

Finally, as far as helping affected communities who are adversely affected by transboundary haze, ASEAN states should now seriously draw some kind of medical preparedness plans to deal with health problems caused by severe impact of the haze. As mentioned earlier, severe haze can raise PSI levels to the hazardous range, causing very significant impacts on mortality as well as respiratory and cardiovascular diseases. It is therefore surprising why outbreaks of haze-related illnesses are not considered as health crises, with regional implications.[18]

Since transboundary haze is likely to be a long-term problem in Southeast Asia, systematic health preparedness measures should become a critical part of transboundary haze risk management in the region. With the help of the WHO, ASEAN could introduce standard government response guidelines to long-term and severe exposure to haze. Response guidelines and PSI indicator levels can help governments to implement measures such as the activation of local evacuation plans based on the PSI, as well as the designation of evacuation centers and the identification of vulnerable groups which require immediate shelter. In this regard, ASEAN can do its part by establishing regional health preparedness guidelines for responses at different PSI levels. A blueprint on the emergency health response measures will be helpful for member states as well as provinces in affected regions. Medical preparedness in the region should also be incorporated into national and local emergency crisis preparedness plans, on top of firefighting systems and forest fire prevention management.

In conclusion, comprehensively addressing the complex problem of transboundary haze pollution is an agenda that ASEAN must decisively act on. If ASEAN were to meaningfully move forward in its goal of establishing an ASEAN Community that is not only peaceful and prosperous, but also "caring and sharing" and indeed people-centered, dealing with the haze problem would thus require both the political will and the commitment of all stakeholders in ASEAN.

---

[18] At the height of the 2015 haze, the number of asthma cases in some areas of Malaysia doubled. O'Callaghan, J. (2015, June 24). Singapore, Malaysia face economic hit from prolonged smog. *Reuters*. Retrieved http://www.reuters.com/article/us-southeastasia-haze-impact-idUSBRE95N0BS20130624

3 Why the Southeast Asian Haze Problem Is Difficult to Solve*

Parkash Chander

Professor and Executive Director, Center for Environmental Economics and Climate Change, Jindal School of Government and Public Policy
parchander@gmail.com
www.parkashchander.com

The Southeast Asian haze is a major environmental problem facing the region and a much studied subject. Thus it may seem that there cannot be much new to say or write about it. Undoubtedly, significant progress has been made in both understanding and instituting policies and mechanisms to tackle it. But the problem continues to persist even after nearly two decades since the region experienced one of the worst episodes in 1997–98. It seems to appear and disappear much like the proverbial Cheshire cat in Alice's Adventures in Wonderland.

The proximate causes for the haze are clear. Indisputably, it stems from the use of fire to burn agricultural residue, clear forest, or prepare land for large plantations and small-scale farms, mainly in Sumatra and Kalimantan of Indonesia and some parts of Malaysia. Fires also occur from vandalism and accidental ignitions in these regions. The purpose of this note is to discuss some obstacles that, unless overcome, may prevent a permanent and effective solution.

To begin with, a common and popular description of the haze is that it is a regional environmental problem. But this is not really an accurate

---

*This note was completed during my visit to Nanyang Technological University (NTU) in Fall 2016. I wish to thank Department of Economics, NTU, for its hospitality and stimulating environment.

description and amounts to looking at the problem mainly from the viewpoint of Singapore and Malaysia that are affected by haze only in El Niño years. This is implicit in the fact that a number of studies estimate damages from haze in the El Niño years, but none for the normal years when the haze remains more or less confined to Indonesia, and affects only Indonesia. A more accurate description might be that haze is mainly a problem for Indonesia that spills over to Singapore and Malaysia in El Niño years.

Why the haze spills over to Singapore and Malaysia also has been only half explained in the literature. It has been claimed that El Niño conditions promote draught and enhance fire activity in the region, either because of drier conditions allowing fires to escape and burn out of control or because farmers take advantage of dry weather and clear more land than usual. But it also needs to be examined whether it is not so much the drier conditions but the direction of winds during El Niño years that causes haze in Singapore and Malaysia. If it is the latter then damages from haze are significantly higher in El Niño years for Singapore and Malaysia, but not for Indonesia. To put it another way, there seems to exist an informational gap regarding the pattern of damages from haze in El Niño vis-à-vis normal years. Filling this gap is important because it can shift the argument for Indonesia from preventing and controlling fires for the sake of its neighbors to preventing and controlling them for the sake of its own people. If the damages for Indonesia herself are significantly higher in El Niño years — and preliminary enquiries show that they are — then it would be easier to convince a better informed Indonesia to make extra efforts to prevent and control fires in the El Niño years for the sake of its own people and not just because it would benefit the neighbors. In fact, highlighting the ill effects of haze only on Singapore and Malaysia can be counterproductive. For instance, Indonesian Vice-President Jusuf Kalla was reported to have chided Malaysia for overreacting about the haze. Instead, a clear recognition of the fact that people in all three countries suffer from haze can lead to empathy and cooperation.

Fires in Indonesia can be divided into two broad categories: (1) fires to clear forest and prepare land for large commercial plantations, and (2) fires to burn agricultural residue and prepare land for small-scale farms. Of these, the fires in the first category may be relatively easier to prevent and control. This is because it is now possible, due to advances in satellite technologies, to show what is happening on the ground. The availability of online monitoring platforms such as the Global Forest Watch-Fires makes it possible to detect and document unsustainable land clearings.[1] Since a number of major companies trading in commodities are headquartered in Singapore and often avail themselves to the financial services and loans from Singapore-based banks, some experts have suggested that, given the information available from the online monitoring platforms, Singapore can effectively integrate its environmental concerns into its policies for investment and finance. However, as is often the case, the deterrence effect of such integrated policies may not be long lasting as the companies may relocate and/or be able to get finance from sources in other countries that are not affected by haze and may not impose similar conditions for loans and investments. Worse, they may be taken over or merged with companies not headquartered in Singapore as they may no longer be able to compete with them. Furthermore, implementation of such integrated policies depends on availability of information regarding ownership of plantations in Indonesia and Malaysia. But both countries have cited sovereignty and legal concerns and refused to share this information.

Singapore is also in an advantageous position for implementing policies that incentivize green practices. This is because it has a 20-percent share of global trade in agricultural commodities and is the world's largest rubber trading hub. Much of the region's supply of palm oil and pulp and paper are traded through Singapore. Thus, it is eminently feasible for it to institute market-based incentives to promote sustainable land use. Coupled with the information available from the online monitoring platforms, Singapore can eco-label goods that are produced by following green practices. Since there is ample evidence that consumers are willing to pay more for goods that are environmentally friendly, eco-labeled goods can be sold at higher prices not only in countries that are affected by haze but also in other countries because of a general concern for the environment among consumers world-wide. The possibility of higher profits facilitated by eco-labelling can provide stronger incentives to at least large plantations in Indonesia and Malaysia to follow green practices and seek

---

[1] The Singapore government, for many years, has been using similar satellite technology to help monitor the situation.

eco-labelling of their products from Singapore. Since the number of environmentally concerned consumers is globally very large, eco-labelling may work much better than calls to boycott haze-linked products by consumers in haze affected countries, especially since, as noted earlier, it is difficult to identify culprit plantations whose products are to be boycotted.

Some estimates show that fires on small-scale farms contribute more than 40 percent of the haze. Therefore, haze cannot be completely prevented without controlling and preventing fires on small-scale farms. But this is a far more difficult task for the following reasons. The small-scale farms are mostly owned by farming families who have, for generations, used fire to burn agricultural residue and clear land. It is difficult to convince them not only of the harms of this traditional practice but also of benefits of green practices. In addition, these farmers are poor and switching to more costly green practices can endanger their economic survival.

Experience from another developing country, namely, India is humbling. In the mainly agricultural province of Punjab in Northwest India, small farmers clear and prepare land for the next crop by burning agricultural residue from the preceding crop. This is a cheap and popular method that has been in practice for a long time. The smoke generated by the burnings drifts downwind and causes haze in densely populated cities. The farmers are small and poor. They neither understand the harms of this practice nor can they afford to follow more expensive green practices. Though the practice has been banned and illegal, it continues unabated. Since the farmers are small and poor and cannot afford to switch to more costly green practices, the practice is not viewed with disapproval by the communities they live in. The government authorities are also reluctant to impose fines for violations as they realize that the very survival of the poor farmers depends on this practice and thus imposing fines is not morally correct. In fact, government authorities are often more than willing to ignore violations of the ban. Significantly, the practice was banned and made illegal by a court, not the government, as no elected government wants to appear to be working against the interests of small and poor farmers, given their large number. In these circumstances, only subsidies can be an effective way to induce the small and poor farmers to adopt green practices.

The same logic more or less applies to the small-scale farmers in Indonesia. Fortunately, Indonesia has received in-principle approval of up to 1 billion US dollars from Norway under the United Nations program called Reducing Emissions from Deforestation and Forest Degradation (REDD). The funds from this program can be used to educate and provide subsidies for adopting green practices by small-scale farmers in the Indonesian provinces. The subsidies may be given in kind as subsidized services of publicly owned equipment and machinery for clearing land without fire. This policy, if successfully implemented, can turn the small-scale farmers into allies in the fight against haze.

4 Transboundary Haze: The Investor Dimension Perspective from Malaysia

Evelyn S. Devadason

Associate Professor, Department of Economics,
Faculty of Economics and Administration, University of Malaya
evelyns@um.edu.my

Introduction

The transboundary pollution[1] or haze from illegal forest fires in Sumatra and Kalimantan is a perennial or cyclical problem for Malaysia, among other affected countries in the region. Significant transboundary haze pollution occurred in 1997/98, 2002, 2005, 2006 and 2013. The cause is large-scale forest clearance for plantations through burning.[2] Critical to this issue is, who are the culprits, and who should be responsible for this large-scale burning?

In 1997 and 1998, when the burning had reached critical levels, plantation companies were considered to have contributed to 34 percent of the burnt areas in Indonesia (Jones, 2006; see also Tacconi *et al.*, 2008; Atkinson, 2014; Jerger, 2014). Oil palm companies, more specifically, were considered the culprit for setting fires to gain land use permits, within the boundary of oil palm concessions (Sheil *et al.*, 2009). Following from the 2006 haze episode, Indonesia alleged that Malaysia- (and Singapore-)

---

[1] Transboundary pollution occurs when a potentially harmful environmental agent is released in one jurisdiction (the source state) and physically migrates through a natural medium, such as air, to another jurisdiction (the affected state) (Heilmann, 2015).

[2] Burning is widely regarded as the quickest and cheapest method to clear high volume forests and forests on peat soil for new plantation sites.

owned palm oil companies were responsible for the fires leading to the haze (Malaysia Today, 2013). For the 2013 haze episode, the World Resource Institute released information that more than three times as many fire hot-spots were observed within logging, pulpwood and oil palm concession areas as compared to outside those concession areas. Approximately 61 percent of the burnt areas within Riau in 2013 were company concessions (Lee *et al.*, 2016). For the same year, the Indonesian Environment Minister released information on eight Malaysian companies with fire hotspots in their concessions, while the companies denied any wrong-doing (Malaysia Today, 2013). As these burnt areas were also illegally occupied by local communities, attribution of fire-events becomes less clear-cut (Lee *et al.*, 2016). Conversely, in 2014, 51 percent of fire alerts were identified as being outside palm oil and logging concessions (cited from Earth Security Group, 2015).

The accountability of plantation companies (oil palm businesses) for the burning and the haze therefore continues to be debated and disputed on grounds that, most fire alerts originate outside concessions, there are other sources[3] involved in the land clearance in Indonesia, apart from plantation companies, and that it is difficult to identify the burning attributable to different sources. Notwithstanding that, attention to palm oil as a plantation commodity linked to haze has increased rapidly (Forsyth, 2014).

Since Malaysia is the biggest investor in the Indonesian oil palm plantation sector, this chapter focuses on the investor dimension, by contextualizing the relevance and role of Malaysian companies in the transboundary haze problem. The intention however is not to maximize the role of Malaysian companies in the transboundary haze, or accord sole responsibility to the plantation sector. This chapter is structured as follows. The second section reviews the state of global players in transboundary pollution to frame the role of investors in the specific problem of transboundary haze. The third section then defines the role of investors through the lens of Malaysia based on what is known and reported about the causes of fires and haze, and Malaysia's own

---

[3] Apart from plantation companies (oil palm businesses), the haze is also attributed to different sources in Indonesia, including migrants, shifting cultivators and small farmers (Jones, 2006).

position and outlook on this issue. The fourth section deliberates on the fractured responses based on national-level actions vis-à-vis the role of the Association of Southeast Asian Nations (ASEAN) in addressing transboundary haze. The fifth section concludes.

Theoretical foundations: Transboundary pollution and strategic players

To provide a conceptual background for this discussion, we review the links between global players (foreign investors) and the environment. The links discussed are based on alternate arguments to the neoliberal assumption that foreign investment is uniformly good for the host state and will result in economic development.

The classic argument of the pollution-haven hypothesis links international investment and environmental pollution (Deng and Xu, 2015). It explains that companies will move their operations to less developed countries to take advantage of less stringent environmental regulations relative to their home country and to save on pollution abatement costs. This hypothesis, which considers environmental regulation and not factor endowment as a pull factor for international investments, has led to the excessive focus on site-specific micro environmental effects on the host economy; indeed a narrow and dated view of the interactions between foreign direct investment (FDI) and the environment.

Instead, the current flow of global capital has resulted in far reaching macro-level implications related to the environmental practices of investors, as FDI can fuel economic activity at a scale and pace that overwhelms host country's regulatory capacity. This strand of literature is dominated by critics of transnational corporations (TNCs) that have become more powerful and less accountable, as they indulge in destructive activities of specific industries, mainly in developing host economies. The critics argue that TNCs in those economies skirt environmental rules, and rely on state officials for access to sites and protection from prosecution (Clapp and Dauvergne, 2008). The dynamics of the interactions between the strategic players (foreign investors) with various groups at the national level have therefore contributed to the complexity of the political economy of environment and investment.

Adding to the complexity is the expanded role of capital flows, which is now linked to environmental effects that go beyond the boundaries of the host economy. In this regard, the international environmental law has evolved to establish legal principles to make states accountable for activities within their jurisdiction, particularly so that those activities do not harm the environment of other states (Jerger, 2014; Laely *et al.*, 2015; Redgewell, 2015; Lee *et al.*, 2016). In relation to the transboundary haze problem, Quah and Johnston (2001) and Quah (2002) point out that what is established in theory is difficult to adopt in practice. Their reasoning is that affluent states such as Singapore, would normally be reluctant to demand for interstate liability against an impoverished culprit state like Indonesia. It could also be problematic to implement the state responsibility principle in the ASEAN context, as state sovereignty is strongly embedded in the ASEAN Charter (Ruland, 2011; Laely *et al.*, 2015). Heilmann (2015) argues otherwise, stating that the severely affected states in the transboundary haze (namely Malaysia and Singapore) have never asked Indonesia to bear state responsibility for the breach of international obligations to control its fires and to incur international liability for the environmental damage because of their own political interests in their oil palm companies in Indonesia.

Notwithstanding the reasons for not invoking the law, transboundary pollution has now gained more recognition from the region as a co-owned environmental problem (Rosenberg, 1999), more so since it is caused, to some degree, by investment linkages. Clearly, what comes out from this regionalized haze problem are: (i) The states and investors need to look beyond the narrow pursuit of their political and economic interests, and acknowledge the transnational effects of the haze (Quah and Johnston, 2002); (ii) The 'culprit' state needs to recognize that investors are drivers of transboundary haze (Lee *et al.*, 2016), and exercise its obligation to regulate and penalize entities for any fire events occurring within its concession boundaries; and (iii) The investors' home country ('culprit' and 'victim') should also take some responsibility for the actions of their companies in host economies. Worth noting here is that the line between a 'culprit' (source) state and a 'victim' (affected) state has also become somewhat blurred, as Indonesia and Malaysia fall neatly into both categories (Quah and Johnston, 2001; Quah, 2002).

The investor dimension: Defining Malaysia's role and outlook

Malaysia, obviously, is directly implicated in the transboundary haze when the role of investors or TNCs (Pye, 2009; Varkkey, 2013) is brought into the equation. The regional investors here refer to those in the oil palm companies. The strong presence of Malaysian companies in the oil palm industry in Indonesia, either wholly Malaysian-owned or their subsidiaries, has drawn them in the transboundary haze problem. Malaysian investors control about 25 percent of the Indonesian oil palm plantation sector (Hanim and Cecep, 2015). With an estimated 184 Indonesian plantations linked to Malaysian companies, the oil palm land bank of Malaysian companies in Indonesia was estimated at 1.8 million hectares in 2013 (Aidenvironment, 2014).

To appraise the role of Malaysian oil palm companies in Indonesia, it is important to go back briefly in history to understand the strategic move of these companies under the auspices of government support. The Malaysian government offered various tax breaks for its companies to venture abroad since the 1990s. For oil palm, the government negotiated investment guarantees with Indonesia and acquired land in Indonesia for oil palm development. Thus, Indonesia became a focal point for oil palm companies in Malaysia, mainly to reap the benefits of economies of scale given the stagnation in the domestic economy, and to take advantage of the abundant undeveloped land (Jamal, 2003; Teoh, 2012; Aidenvironment, 2014) in Indonesia, such as tropical forests, community land and peatlands. The companies that established upstream operations in Indonesia were mainly government-linked companies (GLCs) and private conglomerates (Table 1), some of which formed joint ventures with local companies. There is therefore relentless support for these companies by the Malaysian government, which did not stop with facilitating entry into Indonesia, but continues with looking after the interests of plantation owners, namely through the Malaysian Palm Oil Association (MPOA), the Malaysian Palm Oil Council (MPOC) and the Association of Oil Palm Plantation Investors of Malaysia in Indonesia (APIMI) (see also Heilmann, 2015).

With extensive support, engagement and influence of the government, the industry regionalized rapidly. Consequently, Malaysia has been displaced by Indonesia as the largest producer of palm oil since 2006

Table 1. Major Malaysian Oil Palm Companies in Indonesia

Company	RSPO	ZB	Listing/ Ownership	Indonesia Planting (Ha)
Sime Darby Bhd.	√	√	KLSE	204,466 (289,000+)
IOI Corporation Bhd.*	√ (2016)	√	KLSE	178,000
Genting Plantations*	√ (2014)	√	KLSE	170,000 (193,000+)
Kuala Lumpur Kepong (KLK) Bhd.	√	√	KLSE	137,483
TSH Holdings Sdn. Bhd.	√	√	KLSE	109,000
Oriental Holdings Bhd.			KLSE	97,000+
Tadmax Resources Bhd., Bumimas Raya Sdn. Bhd., Pacific Inter Link Sdn. Bhd., Yakima Dijaya Sdn. Bhd. (joint-venture)			KLSE	80,000+
Pinehill Pacific Bhd.	√	√	KLSE	73,000+
CB Industrial Product				68,000+
Felda Global Ventures Holdings Bhd.	√	√	KLSE	56,422
Lion Forest Industries Bhd.			KLSE	53,000+
Trurich Resources Sdn. Bhd.			Private	42,000+
Kulim Malaysia Bhd.	√	√	KLSE	41,000+
Glenealy Plantations (Malaya) Bhd.			KLSE	41,000+
Southern Group				39,000+
NPC Resources Bhd.			KLSE	36,000+
TDM	√	√	KLSE	30,000+
Kwantas Corporation Bhd.			KLSE	28,000+
IJM Plantations Bhd.	√	√	KLSE	27,491
Low Yat Group			Private	27,000+
Chellam Plantations Group			Private	25,000
Ahmad Zaki Resources Bhd.			KLSE	21,000+
Chin Teck Plantations Bhd.			KLSE	18,000+
Kumpulan FIMA Bhd.			KLSE	18,000+
MKH Bhd.			KLSE	16,000+

(*Continued*)

Table 1. (*Continued*)

Company	RSPO	ZB	Listing/ Ownership	Indonesia Planting (Ha)
Delloyd Ventures Bhd.			KLSE	16,000+
Golden Land Bhd.			KLSE	16,000+
SADC/PPPNP (Perak State Government)				15,000+
NAFAS			Private	14,000+
QL Resources Bhd.			KLSE	10,000+
United Plantations Bhd.	√	√	KLSE	9,815 (35,000+)
Tabung Haji Plantations (THP)	X	√	KLSE	8,500 (84,000+)
Southern Acids (M) Bhd.			KLSE	8,000+
Samling Group			Private	6,085
Batu Kawan Bhd.			KLSE	6,000+
Cepatwawasan Group Bhd.			KLSE	5,000+

Notes: RSPO — Roundtable Sustainable Palm Oil; ZB — zero burning; * — suspended as member of RSPO; X — scrapped in 2013; + — land banks (ha).
Sources: (1) Earth Security Group (2015); (2) Aidenvironment (2014); (3) Annual reports of companies.

(Figure 1), due to the limitation of land for further expansion in the home economy. It is an export-oriented industry that contributes significantly to the economic development and revenue of Malaysia, but involves oligarchy controls over production and trade. There are therefore definitely entrenched interests in palm oil companies. Unfortunately, their expansions in Indonesia come at a cost for the environment, and are making this a controversial industry, touching also on national sensitivity.

The palm oil sector has come under scrutiny of the international environmental community for issues related to non-sustainable development. Based on the Sustainable Palm Oil Transparency Toolkit (SPOTT), which ranks the world's largest 25 publicly listed palm oil companies in terms of transparency around the environmental performance of their operations, five Malaysian companies garnered the lowest scores and bottomed the list (Butler, 2014). They include Malaysia Airport Holdings Berhad (MAHB), TSH Resources Berhad, Sarawak Oil Palms Berhad, Jaya Tiasa

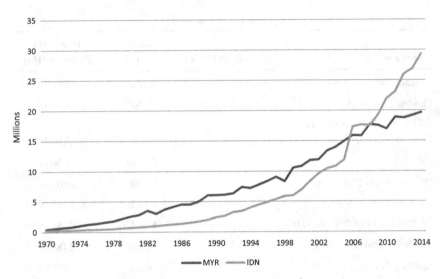

Figure 1. Production of Crude Palm Oil in Malaysia and Indonesia (Million Tonnes)
Note: MYR — Malaysia; IDN — Indonesia.
Source: FAOSTAT.

Holdings Berhad and QL Resources Berhad. Among the five companies, TSH Resources Berhad and QL Resources Berhad have plantations in Indonesia (see Table 1), and the former is a member of the Roundtable on Sustainable Palm Oil (RSPO).[4] Following which, some have claimed that though companies have committed themselves to the RSPO, the use of fire for land clearing is still rampant (Richter, 2009). Things looked bad when major companies like Sime Darby Berhad and Kuala Lumpur Kepong (KLK) Berhad were alleged by the Indonesian government to have contributed to the 2013 haze. On grounds that RSPO has no binding enforcement instrument, among other reasons, some non-governmental organizations (NGOs) reject this association (see also Teoh, 2012). The recent signing of the memorandum of understanding (MOU) to form a new cartel-like establishment called Council of Palm Oil Producing Countries (CPOPC) between Malaysia and Indonesia in 2015, which will effectively control 85 percent of the world's oil production, also raises

---

[4]The RSPO is a multi-stakeholder body and an international association founded in 2004. It is the main global certification body for sustainable palm oil.

concerns that commitments made at the RSPO will not be heeded. If that were the case, is the certification of sustainable palm oil production sufficient to address unethical conduct of land clearance through burning?

It is also noted from Table 1 that most of the major companies in Malaysia have committed to zero burning.[5] The focus on zero-burning approach or a blanket ban on burning is considered purely a distraction from the main sources of haze pollution (Murdiyarso *et al.*, 2004; Tacconi *et al.*, 2008). Tacconi argues that the focus should be more realistic given existing practices in land management and pressures for development, and therefore zero burning should only be promoted for specific locations that pose risks to haze pollution (see also Quah, 2002). The policy may also be counterproductive if it is used as a leverage by companies to deny any allegations made against them for the burning issue. For example, in 2013, when Indonesia announced that they were going to charge PT Adei Plantation (a subsidiary of Malaysia's KLK) and TH Plantations Berhad (a subsidiary for Malaysia's Tabung Haji Plantations (THP)) for environmental damage, KLK and THP denied the involvement of their respective subsidiaries of any wrong-doings on grounds that they observe a zero-burning policy, among other reasons. Given the bad track record of PT Adei Plantation, which was charged for open burning in 2001 by the Indonesia authorities, it became even more difficult to accept KLK's claims (SAM, 2014). The concern here is that the zero-burning policy should not be a 'greenwash'[6] or an excuse for investigations to be halted on potential violators.

While some continue to argue that the role of Malaysian oil palm companies in the burning misconduct still remains unclear, the environmental backlash for these companies, which began after the 1997 haze episode, continues to intensify. Two recent examples of consumer backlash that have affected big players in the Malaysian oil palm industry include the following. First, the removal of Asian companies, which pose environmental damage by turning rainforest in Indonesia into palm oil plantations, by Norway's sovereign wealth fund from its investment portfolio. Two of which are

---

[5]The zero burning policy was adopted at the 6th ASEAN Ministerial Meeting on Haze in 1999, with a view to promote its application by plantation companies (and timber firms).
[6]'Greenwash' refers to a phenomenon where a company tries to convince the community that it is environmentally responsible, while the purpose is more about image than substance (Clapp and Dauvergne, 2008).

Malaysian companies that have palm oil holdings in Indonesia, conglomerate Genting Berhad and construction firm IJM Corporation Berhad (Kozlowska, 2015). Second, the removal of IOI Group as a supplier by three of the world's top food and consumer goods companies, Unilever, Mars and Kellogg's, because of evidence of deforestation of the company's plantation in Indonesia. The decision by the three global giants came soon after the IOI Group was suspended from the RSPO (Fogarty, 2016).

How has Malaysia responded to the allegations that Malaysian oil palm plantation firms are responsible for the fires in Indonesia to clear land? Varkkey (2013) argues that allegations against these firms are largely disregarded, and their actions defended by the Malaysian government because the companies, which include GLCs and non-GLCs, have close patronage relationships and links with government elites. In that case, industry-level actions with non-interference from the state seem remote, given the tight link between the two. Evidently, patronage relationships, as argued by Varkkey, also co-exist in Indonesia[7] (see also Murdiyarso *et al.*, 2004; Weatherbee, 2007; Javadi and Han, 2015), as offenders that are well-connected to the political and business elites in the country are rarely dealt with. Atkinson (2014), supports this argument as he too cites crony relationships, between Malaysian (and Singaporean) companies with provincial and district officials (Cotton, 1999; Mabey and McNally, 1999; Jamal, 2003; Varkkey, 2012; Sunchindah, 2015), for creating a *de facto* ban on enforcement actions against elite groups, or leniency on legal action related to fire producing haze (Varkkey, 2016). It appears that the interactions between foreign investors and the local regulator in both the home and host economies have accorded investors some leveraging power in the haze-related issue.

Malaysia has therefore yet to come up with any decisive action despite Indonesia's call for affected states to check their companies. This is not surprising, as Malaysia's responses, historically, have not always been impressive, when it involves the haze problem. In the 1990s, the local media was suppressed from reporting on the haze, on grounds that it was smearing the image of the country (Cotton, 1999). It was not until the late

---

[7] Butt (2011) provides a detailed account of legal uncertainties amidst the dramatic rise in law-making in a decentralized Indonesia that have resulted in rampant corruption, especially in investment related matters. It has provided a fertile ground for illicit payments required to commence and run investments, and to resolve disputes arising from them.

1990s that Malaysia began to accept some responsibility to the transboundary haze. The then Prime Minister of Malaysia, Dr. Mahathir bin Mohamad, made several statements of mutual responsibility with Indonesia for the haze issue (Forsyth, 2014). Following which, Malaysia affirmed that action would be taken again companies guilty of the burning. However, they did not follow-through with any conclusive action. It seems that statements of intent are expressed during and immediately after the haze situations, and when the crisis wanes, the haze issue goes into a silent mode and is evaded in the political agenda.

Interestingly, the response of Malaysia towards the haze issue, particularly since 2013, seems to have changed. Forsyth (2014) observes that most notions of blame in the Malaysian media were expressed against Indonesia. Likewise, the Indonesian media also made more references to palm oil for the 2013 haze. Prime Minister Mohammad Najib bin Tun Haji Abdul Razak responded to this by shifting the responsibility to Indonesia to gather evidence and take action against the companies concerned (Hunter, 2015). However, as the Chair of ASEAN in 2015, Malaysia had to assume some role and thereby addressed the issue of the haze pollution in the context of the 'people-centred and people-oriented' rights to clean air at the 36th General Assembly of the ASEAN Inter Parliamentary Assembly held in Kuala Lumpur. Now, both countries have decided to cooperate and set green standards to produce environmentally sustainable palm oil, and to harmonize the principles and criteria used for their respective certification of sustainable palm oil under the newly-formed CPOPC. At this juncture, the new regulatory framework and the sustainability standards of the CPOPC remain mired in ambiguity.

It is clearly in the interest of the states and oil palm companies to ensure sustainability, as the controversy over the role of the latter in the burning issue in Indonesia will continue and maybe even intensify. Joint efforts are therefore needed as Malaysia cannot tackle these problems alone. The haze source, in this case Indonesia, also needs to take seriously sustainable palm oil production in its own courtyard.

Regional actions vs. national responses: Fractured responses

Since the haze implicates and affects several countries in the region, ASEAN seemed a natural forum in which to address the problem. The ASEAN

Agreement on Transboundary Haze Pollution (AATHP),[8] which was signed in June 2002 and came into force in 2003, was consequently borne out of the need to have a common set of laws at the regional level to address environmental issues. It is the first instrument to tackle environmental issues such as transboundary pollution within ASEAN. Worth mentioning here is that Indonesia, the most critical actor in the haze problem, only ratified the Agreement in September 2014 and became the last country to join the Agreement.

Unfortunately, the AATHP has weak mechanisms that are unlikely to bring about change (Heilmann, 2015). First, the Agreement lacks enforceable mandatory provisions (see Jones, 2006; Jones and Smith, 2007; Laely *et al.*, 2015; Sunchindah, 2015). As it does not come with provisions for compliance, among others, the AATHP ranks low based on its theoretical strengths relative to other international environmental agreements (Tacconi *et al.*, 2008). This is clearly a downside to the Agreement as companies will only comply if authorities implement laws strictly and impose penalties. Second, the focus of ASEAN through the AATHP has largely been on technical cooperation (such as deployment of fire-fighting forces, sharing of technology and enforcement approaches), to assist Indonesia in its fire management needs. This however does not provide a solution for the problem.[9] While the move by ASEAN in 2013 to adopt the joint monitoring system and share satellite data to help better locate fire hotspots is considered a positive development in deriving solutions to the problem, it is still regarded as insufficient to identify the culprits for the burning. Technical information is needed, such as the disclosure of a single accurate consistent concession map[10] by Indonesia, to establish if these hotspots are on land owned by plantation companies. Then again, it is argued that this information is still inadequate to address actors responsible for starting the fire.

---

[8] It contains provisions on monitoring, assessment and prevention, technical cooperation and scientific research, mechanisms for coordination, lines of communication, and simplified customs and immigration procedures for disaster relief.

[9] Haze is a problem that would be more effectively addressed through prevention than remediation (Murdiyarso *et al.*, 2004; Atkinson, 2014).

[10] Approximately 10.2 percent of industrial concessions from oil palm, logging, pulpwood, and coal mining industries in Indonesia overlap (Lee *et al.*, 2016).

From a regional perspective, while transboundary haze gained the attention of ASEAN since the 1990s, the efforts taken by ASEAN to date has been largely ineffective in stopping the haze from recurring, even with the AATHP. Apart from the fact that the AATHP appears to invoke the ASEAN way of non-interference in handling regional matters (Tacconi *et al.*, 2008), others argue that there is also a lack of political will (Jones, 2014) at the regional front, to follow-through with this problem. Following which, some conclude that the transboundary haze reflects the failure of regional governance (Tay, 2002; Aggarwal and Chow, 2010; Nguitragool, 2011). The debate for tackling transboundary haze has therefore gradually shifted from mere regional cooperation, to national-level reforms within Indonesia, coupled with unilateral actions from affected states.

It is recognized that national responses, particularly from the source and affected (namely culprit) states, are needed to complement regional cooperation given that the AATHP has no route for direct legal redress (Heilmann, 2015). Progress in Indonesia to combat transboundary haze however remains weak due to the lack of political will, legal reform and administrative coordination (Jones, 2006; Mayer 2006; Hanim and Cecep, 2015). Though there were indications of growing political will under the previous President Susilo Bambang Yudhoyono, recalcitrants persist in some segments of the Indonesian government. For example, deficiencies in institutional capacity in terms of division of responsibilities to different public organizations (central, provincial and district) with overlapping functions (see also Butt, 2011) in Indonesia have affected the coordination and implementation of measures to prevent and mitigate the fires (Sunchindah, 2015). It seems poor governance is also a feature of national-level actions.

In contrast to Malaysia, Singapore, an affected state which has also been implicated as a contributor to the haze, has taken on a bolder unilateral approach to the transboundary issue with the enactment of its own 2014 Singapore Transboundary Haze Act.[11] The Act makes it an offence for Singaporean companies to engage in any conduct that causes haze pollution, even if the activities of the companies originate outside its political

---

[11] Effective implementation of criminal liability actions against entities responsible for causing or contributing to the haze pollution under the Act, though problematic (see Lee *et al.*, 2016), can direct appropriate attention to the perpetrators (Sunchindah, 2015).

boundaries. Similar legal options to punish companies responsible for the haze-causing fires in Indonesia remain debated in the Malaysian case. While some forward their views that a similar move like Singapore for Malaysia will only detract Indonesia from enforcing its own laws, others opine that it would instead exert pressure on Indonesia to take on substantial action. However, even for the Singapore laws to be effective, they again need to be complemented with better enforcement of anti-burning laws and other reforms, specifically better control and transparency of land ownership, concessions and classifications in Indonesia.

Clearly, the transboundary haze problem involves various actors and requires a confluence of actions and cooperation at the national and regional levels. What is obvious is that both national and regional responses have been somewhat reactive. There is still lack of proactive responses, which at best should come from Indonesia and Malaysia. If legal action is rightfully not taken against the TNCs in the host state where the environmental laws have been breached, nor is the parent company's activities regulated in the home state, then there will be no accountability of these actors in the haze problem.

Concluding remarks

The Malaysian palm oil industry has expanded spatially within the region, largely driven by TNCs. Their large and widespread presence in Indonesia, coupled with mounting allegations of fires to clear concession areas, evidence of bad environmental practices, suspension of big companies from the RSPO, holds them somewhat accountable for the haze causing fires in Indonesia. The transnational or investor linkages, however, are only half the story for understanding the key actors or strategic players involved in the transboundary pollution problem. Underlying these investor connections, more importantly, are the collusion between the investors (palm oil companies) with local authorities in the host economy (Indonesia) and also the patronage networks in the home economy (Malaysia). Therefore, the political constructs in this environmental problem, linked to the investor dimension, become a defining problem in the combat of transboundary haze. Namely, when the symbiosis between investors and states has turned into a bad thing, leading to undesirable outcomes, all other efforts at the national and regional levels to combat the problem will remain, 'hazy', 'fractured' and limited, and not go beyond rhetoric.

References

Aggarwal, V. K. and Chow, J. T. (2010). The perils of consensus: How ASEAN's meta-regime undermines economic and environmental cooperation, *Review of International Political Economy*, 17: 262–290.

Aidenvironment (2014). Malaysian overseas foreign direct investment in oil palm land bank: Scale and sustainability impact. Commissioned by Sahabat Alam Malaysia (SAM). Retrieved http://www.aidenvironment.org/media/uploads/documents/170603_3093_SAM_OFDI_17_June_2014.pdf

Atkinson, C. L. (2014). Deforestation and transboundary haze in Indonesia: Path dependence and elite influences, *Environment and Urbanization ASIA*, 5(2): 253–267.

Butler, R. A. (2014, November 20). Ranking the world's best — and worst — palm oil companies in terms of sustainability, *Mongabay*. Retrieved http://news.mongabay.com/2014/11/ranking-the-worlds-best-and-worst-palm-oil-companies-in-terms-of-sustainability/

Butt, S. (2011). Foreign investment in Indonesia: The problem of legal uncertainty, in Bath, V. and Nottage, L. (eds.), *Foreign Investment and Dispute Resolution Law and Practice in Asia*, Routledge: London and New York, pp. 112–134.

Clapp, J. and Dauvergne, P. (2008). Global investment and the environment, in *Paths to a Green World: The Political Economy of the Global Environment*, Academic Foundation: New Delhi, pp.157–188.

Cotton, J. (1999). The "haze" over Southeast Asia: Challenging the ASEAN mode of regional engagement, *Pacific Affairs*, 72(3): 331–351.

Deng, Y. and Xu, H. (2015). International direct investment and transboundary pollution: An empirical analysis of complex networks, *Sustainability*, 7: 3933–3957.

Earth Security Group (2015). *The Earth Security Index 2015: Managing Global Resource Risks and Resilience in the 21st Century*, Earth Security Group: London.

Fogarty, D. (2016, April 7). 3 global giants drop Malaysian palm oil supplier over deforestation. *The Straits Times*. Retrieved http://www.asianews.network/content/3-global-giants-drop-malaysian-palm-oil-supplier-over-deforestation-13787

Forsyth, T. (2014). Public concerns about transboundary haze: A comparison of Indonesia, Singapore and Malaysia, *Global Environmental Change*, 25: 76–86.

Hanim, K. and Cecep A. (2015). Transboundary haze polluters and accountability: The legal landscape in Indonesia and Malaysia, paper presented at the 18th International Academic Conference, London, 25 August.

Heilmann, D. (2015). After Indonesia's ratification: The ASEAN agreement on transboundary haze pollution and its effectiveness as regional environmental governance tool, *Southeast Asian Affairs*, 34(3): 95–121.

Hunter, M. (2015, November 4). Opinion: Malaysia needs to fix its own haze problem first. *Asian Correspondent*. Retrieved https://asiancorrespondent.com/2015/11/malaysia-needs-to-look-at-its-own-sources-of-haze-generation-first/

Jamal O. (2003). Linking agricultural trade, land demand, and environment externalities: Case of oil palm in Southeast Asia, *ASEAN Economic Bulletin*, 20(3): 244–255.

Javadi, T. and Han, D. (2015, October 9). Clearing the diplomatic haze: Indonesia's tense ties with Singapore and Malaysia, RSIS Commentary No. 211, S. Rajaratnam School of International Studies (RSIS), Nanyang Technological University: Singapore.

Jerger, D. B. Jr. (2014). Indonesia's role in realizing the goals of ASEAN's Agreement on Transboundary Haze Pollution, *Sustainable Development Law & Policy*, 14(1): 35–45.

Jones, D. S. (2006). ASEAN and transboundary haze pollution in Southeast Asia, *Asia-Europe Journal*, 4: 431–446.

Jones, W. J. (2014). Human security & ASEAN transboundary haze: An idea that never came, *Journal of Alternative Perspectives in Social Sciences*, 5(4): 603–623.

Jones, D. M. and Smith, M. L. R. (2007). Making process, not progress: ASEAN and the evolving East Asian regional order, *International Security*, 32(1): 148–184.

Kozlowska, H. (2015, August 17). Norway's giant wealth fund decides that palm oil is terrible. *Quartz*. Retrieved http://qz.com/481186/norways-giant-wealth-fund-decides-that-palm-oil-is-terrible/

Laely, N., Shawkat, A. and Zada, L. (2015). The influence of international law upon ASEAN approaches in addressing transboundary haze pollution in Southeast Asia, *Contemporary Southeast Asia: A Journal of International and Strategic Affairs*, 37(2): 183–210.

Lee, J. S. H., Jaafar, Z., Tan, A. K. J., Carrasco, L. R., Ewing. J. J., Bickford, D. P., Webb, E. L. and Lian, P. K. (2016). Toward clearer skies: Challenges in regulating transboundary haze in Southeast Asia, *Environmental Science & Policy*, 55: 87–95.

Mabey, N. and McNally, R. (1999). Foreign direct investment and the environment: From pollution havens to sustainable development, World Wildlife

Fund (WWF): United Kingdom. Retrieved http://www.oecd.org/investment/mne/2089912.pdf

Malaysia Today. (2013, June 22). Haze update: Indonesia names eight companies investigated for burning. *The Straits Times*. Retrieved http://www.malaysia-today.net/haze-update-indonesia-names-eight-companies-investigated-for-burning/

Mayer, J. (2006). Transboundary perspectives on managing Indonesia's fires, *Journal of Environment and Development*, 15(2): 202–223.

Murdiyarso, D., Lebel, L., Gintings, A. N., Tampubolon, S. M. H., Heil, A. and Wasson, M. (2004). Policy responses to complex environmental problems: Insights from science-policy activity on transboundary haze from vegetation fires in Southeast Asia, *Agriculture, Ecosystems and Environment*, 104: 47–56.

Nguitragool, P. (2011). Negotiating the haze treaty, *Asian Survey*, 51: 356–378.

Pye, O. (2009). Palm oil as a transnational crisis in South-East Asia, *Austrian Journal of Southeast Asian Studies*, 2(2): 81–101.

Quah, E. (2002). Transboundary pollution in Southeast Asia: The Indonesian fires, *World Development*, 30(3): 429–441.

Quah, E. and Johnston, D. (2001). Forest fires and environmental haze in Southeast Asia: Using the 'stakeholder' approach to assign costs and responsibilities, *Journal of Environmental Management*, 63: 181–191.

Redgewell, C. (2015). Transboundary pollution: Principles, policy and practice, in Jayakumar, S., Koh, T., Beckman, R., Phan, H. D. (eds.), *Transboundary Pollution. Evolving Issues of International Law and Policy*, Edward Elgar Publishing Limited: United Kingdom.

Richter, B. (2009). Environmental challenges and the controversy about palm oil production – Case studies from Malaysia, Indonesia and Myanmar. Retrieved http://library.fes.de/pdf-files/iez/06769.pdf

Rosenberg, D. (1999). Environmental pollution around the South China Sea: Developing a regional response, *Contemporary Southeast Asia*, 21(1): 119–145.

Ruland, J. (2011). Southeast Asian regionalism and global governance: "multilateral utility" or "hedging utility", *Contemporary Southeast Asia*, 33(1): 83–112.

Sheil, D., Casson, A., Meijaard, E., van Noordwijk, M., Gaskell, J., Sunderland-Groves, J., Wertz, K. and Kanninen, M. (2009). The impacts and opportunities of oil palm in Southeast Asia, Occasional Paper No. 51, Center for International Forestry Research (CIFOR): Indonesia.

Sunchindah, A. (2015). Transboundary haze pollution problem in Southeast Asia: Reframing ASEAN's response, ERIA Discussion Paper Series 2015–82, Economic Research Institute for ASEAN: Jakarta.

Tacconi, L., Jotzo, F. and Grafton, R. Q. (2008). Local causes, regional co-operation and global financing for environmental problems: The case for Southeast Asian pollution, *International Environmental Agreements*, 8: 1–16.

Tay, S. S. C. (2002). Fires and haze in Southeast Asia, in Noda, P. J. (ed.), *Cross-Sectoral Partnerships in Enhancing Human Security*, Japan Center for International Exchange: Tokyo, pp. 53–80.

Teoh C. H. (2012). Malaysian corporations as strategic players in Southeast Asia's palm oil industry, in Pye, O. and Bhattacharya, J. (eds.), *The Palm Oil Controversy in Southeast Asia: A Transnational Perspective*, Institute of Southeast Asian Studies: Singapore, pp. 19–47.

Varkkey, H. (2012). Patronage politics as a driver of economic regionalisation: The Indonesian oil palm sector and transboundary haze, *Asia Pacific Viewpoint*, 53(3): 314–329.

Varkkey, H. (2013). Malaysian investors in the Indonesian oil palm plantation sector: Home state facilitation and transboundary haze, *Asia Pacific Business Review*, 19(3): 381–401.

Varkkey, H. (2016). *The Haze Problem in Southeast Asia: Palm Oil and Patronage*, Routledge: London.

Weatherbee, D. E. (2007). Southeast Asia in 2006: Déjà vu all over again, *Southeast Asian Affairs*, 2007: 3–30.

5 The Role of the Polluter-Pays Principle in Managing the Haze in Southeast Asia

Preston Devasia

Consultant in Biodiversity and Environmental Management, Singapore

The recurrent haze in Southeast Asia, due to burning of forests for clearing the land in Sumatra and Kalimantan in Indonesia, had reached an unmanageable situation with air pollution in Indonesia and surrounding countries including Singapore, Malaysia, Thailand, Brunei, Vietnam, Cambodia and the Philippines leading to health concerns, discomfort and impaired visibility. Why does this happen again and again over many years?

Hardin (1968) in his seminal essay on "The Tragedy of the Commons" argues that a resource which is in the commons is bound to degradation due to poor husbanding and overexploitation. When a school of fish is spotted in the ocean, the fisherman quickly catches all he can as otherwise other fishermen will catch all they can. There is no reward for conservation, as the resource is not managed with clear implementable rules by a managing body or self-organizing principles (Ostrom *et al.*, 1999). However, if the fisherman was farming Tilapia fish in his own pond, there would be an incentive for proper management of the pond with harvesting methods that ensure the fingerlings mature and are replaced when harvested. The property rights of the Tilapia farmer ensure a sustainable management method.

In a finite Earth with an ever increasing population, there is an urgent need to manage the commons. Most of the commons on earth harbor mixed and merit goods. Mixed goods have characteristics of both private goods (rival, excludable internal benefits) and of public goods (external benefits which are non-rival and non-excludable). Merit goods are those which are considered so good that the government provides the quantity that it thinks the customers ought to consume by overriding their preferences. Clearly, conservation of natural resources, including biodiversity, and providing a clean environment for sustainable living and health are both mixed and merit goods (Pearce, 1998). Hence there is a case for governments to take over management of the commons. In a global village, managing the commons is a duty of the governments and in international spaces it has to be through intergovernmental cooperation. There should be an economic cost to avoid misuse of the commons so that the commons can be managed for the greater good.

The Organisation for Economic Co-operation and Development (OECD) (1975) reports: "The Polluter-Pays Principle means that the polluter should bear the expenses of carrying the measures decided by public authorities to ensure that the environment is in an acceptable state. In other words, the cost of those measures should be reflected in the cost of goods and services which cause pollution in production and/or in consumption. Such measures should not be accompanied by subsidies that would create significant distortions in international trade and investment."

The haze poses economic, social and environmental costs. Based on the Polluter-Pays Principle adopted by the Association of South East Asian Nations (ASEAN) countries in 1985 through the ASEAN Agreement on the Conservation of Nature and Natural Resources [Article 10(d)], the polluter should bear the cost. In other words, there should be a Pigouvian tax (Baumol and Oates, 1988) or pollution charge. Since the individual polluters are difficult to identify and extract the cost, it is suggested that the Indonesian government should bear responsibility and pay the pollution charge to the affected countries. Luppi *et al.* (2012) in their article entitled "The Rise and Fall of the Polluter-Pays Principle in Developing Countries" published in the *International Review of Law and Economics*, has suggested that in environments characterized by widespread poverty, high interest rates, judicial delays and uncertainty in adjudication, the concerned government may be directly held responsible for payment and environmental monitoring. It should be the endeavor of the Indonesian government to impose and recover the pollution charge from the landowners responsible for the haze. The quantum of the pollution charge should be worked out by economists.

Acknowledgments

The author is thankful to Professor S. Jayakumar for critical comments on an earlier version of this article. He is further grateful to Estherpwa Theresia Megawati for valuable criticism and support. This article is dedicated to the memory of the late Devasia Koikkara who encouraged the author to apply knowledge acquired to practical use.

References

Association of South East Asian Nations (ASEAN) (1985). ASEAN Agreement on the Conservation of Nature and Natural Resources, Article 10(d). Retrieved http://environment.asean.org/agreement-on-the-conservation-of-nature-and-natural-resources/.

Baumol, W. J. and Oates W. E. (1988). *The Theory of Environmental Policy*, second edition (pp. 21–23). Cambridge University Press: Cambridge.

Hardin, G. (1968). The tragedy of the commons, *Science*, 162: 1243–1248.

Luppi, B., Parisi, F. and Rajagopalan, S. (2012). The rise and fall of the Polluter-Pays Principle in developing countries, *International Review of Law and Economics*, 32: 135–144.

Organisation for Economic Co-operation and Development (OECD). (1975). *The Polluter Pays Principle*, OECD, Director of Information: Paris.

Ostrom, E., Burger, J., Field, C. B., Norgaard, R. B. and Policansky, D. (1999). Revisiting the commons: Local lessons, global challenges, *Science*, 284: 278–282.

Pearce, D. W. (1998). Valuing the environment., in Pearce, D. W. (ed.), *Economics and Environment: Essays on Ecological Economics and Sustainable Development*. Cheltenham, UK: Edward Elgar Publishing Limited, pp. 13–33.

6 Indonesia's Uphill Battle against Dangerous Land Clearance

Jackson Ewing

Director, Asian Sustainability, Asia Society Policy Institute
JEwing@asiasociety.org

In mid-April 2016 Indonesian President Joko 'Jokowi' Widodo announced a planned moratorium on new palm oil concessions. The following July, Jokowi held a ministerial meeting with Environment and Forestry, Trade, Industry, and Spatial Planning Ministers to take the initiative forward — combining these actions with a continuing 2011 moratorium on peatland and forest exploitation and a drive to unify the convoluted concession maps that have plagued Indonesian forest protection for decades.

These measures are occurring in concert with the government putting forward ambitious climate change mitigation goals, developing REDD+ forest protection programs, and cooperating more substantively with regional partners to battle the land-clearing fires causing haze in Southeast Asia.

Jokowi, as arguably was the case with his predecessor Susilo Bambang Yudhoyono, seems to take the environmental, social and political problems of land clearance and estate farming in Indonesia seriously, and to be prioritizing their redress.

The Yudhoyono administration expanded programs on community-based forest management and fire prevention, levied a fine of over 25 million US dollars on an offending palm oil producer, and arrested executives of companies allegedly behind the current fires.

Regional neighbors meanwhile have contributed resources and prioritized the Asean Haze Monitoring System (AHMS), which uses digital maps to share information on concession land in forest fire-prone areas. Singapore passed the Transboundary Haze Pollution law in 2014 to hold palm oil and paper companies legally responsible in Singapore for burning activities in elsewhere that affect the island state. Meanwhile major corporate players, including Wilmar, Cargill, Golden Agri-Resources, Asian Agri, Musim Mas, and others, have become increasingly concerned with the reputational and business costs of pollutive land clearance and adopted stronger conservation and human rights policies.

So why have Indonesian citizens and those in neighboring countries experienced the worst transboundary haze episodes on Yudhoyono and Jokowi's watch? And why are these same stakeholders bracing for potentially acute haze episodes to come? Three factors stand out from the rest.

Expanding palm oil and paper sectors

The forces causing the haze are outpacing efforts to mitigate it. While Indonesia has taken recent steps to combat the haze-causing fires, it has simultaneously advanced its palm oil and pulp and paper sectors as key engines of the wider economy. After surpassing Malaysia as the world's leading producer of palm oil in 2006, Indonesia announced plans to double production and brought millions of new hectares under cultivation. Total harvested oil palm area in Indonesia grew from 4.1 million hectares in 2006 to an estimated 8.9 million hectares in 2015 — when Indonesia was producing over 32.5 million tons of crude palm oil and exported 80 percent of it for some 18.6 billion US dollars. The sector will cover some 17 million hectares by 2025 under current trends, which have extended through multiple moratoria on forest and peatland clearing.

Meanwhile demand for pulp and paper continues to rise in emerging Asian economies, particularly China, with Indonesian plantations growing apace. The industry is expanding at around 5 percent annually, and the Indonesian Industry Ministry voiced optimism in February 2016 that the country's pulp and paper production could expand annual production from 7.9 million tons to 8.7 million tons in 2017. It is difficult to see these land-intensive sectors expanding along current trend lines without haze implications.

The boom in these sectors is also changing their structures and characteristics. Expansion has traditionally been defined largely by estate-level land clearance, with blurred lines between corporate firms and the small-scale landowners they often contract to. There is now a growing presence of mid-sized actors that develop plantations but have scant or non-existent public profiles.

These actors gain official and unofficial concessions from local governments, whose leaders seek capital for their budgets, their campaigns, and at times their personal gain. Haze does not present the same reputational risks to these mid-level operatives as it does to large corporations which have greater incentives for implementing haze prevention policies.

The fingerprints of government interests also seem to appallingly appear on the July 2016 dismantling the Indonesia Palm Oil Pledge (IPOP). The IPOP was an entity set up by the leading large palm oil interests and the Indonesia Chamber of Commerce to promote responsible practices. Its architects claim to have dissolved the body under pressure from government actors interested in perpetuating business as usual.

Changing weather

The source areas of the haze are getting hotter and dryer. Burning remains an attractive method for land clearing because it is quick and efficient, requires minimal labor, enriches soils, and provides a default strategy in lieu of affordable alternatives. The worst heatwave conditions in decades plagued Southeast Asia during the 2015–2016 El Niño period; conditions which make fires more likely to become large and difficult to control. In carbon-rich peatlands, these fires can burn for weeks and spread far beyond their areas of origin; which in turn problematizes efforts to establish culpability.

Relationships connecting climate change and El Niño periods are difficult to pin down in all their intricacies, but climate change trends — with each year setting global records for average temperatures — increase the threat of future fires. In equatorial Southeast Asia, dry season months are getting hotter and dryer, while wet season months are becoming more defined by major rain events that are difficult to predict and plan for. These climate disruptions amplify the already formidable regional haze challenge.

Policies playing catch-up

Efforts to address the haze will take years and nascent policies are not appreciably changing the short-term conditions on the ground in Indonesian plantations. It remains difficult to identify haze-causing culprits even with new laws, legislation, greater enforcement ambitions, and better maps detailing where concessions are situated. Time may improve the effectiveness of these mechanisms, but, as recent years and current haze threats demonstrate, they have not yet proven up to the near-term challenge.

Affected Indonesian citizens suffer more painfully from the haze than do their neighbors, and hope for solutions as much as anyone. Such solutions may be taking shape, and the disastrous haze episodes of 2015 have reminded policymakers of the high stakes involved. Doubling-down on regional cooperation is a task without viable alternatives. An ASEAN meeting in August 2016 on combatting the haze was critiqued by some as not holding Indonesia's feet to the fire, but it is through quiet and dogged diplomacy, capacity building, and continuing pressure on the private sector that the haze will ultimately be reined in.

7 Seeing the Big Picture through the Haze

Min Geh

Board Member, BirdLife International (Asia), Singapore
Immediate Past President, Nature Society, Singapore

Agus Budi Utomo

Adviser, Forest and Biodiversity, Burung Indonesia, Bogor, Indonesia
agus@burung.org

Sultana Bashir

Conservation and Sustainability Consultant, United Kingdom
sultana.bashir1@gmail.com

Seeing the big picture

When humans first appeared on Earth we were small in numbers and insignificant in our impact, but in a relatively short time in the Earth's history we have dominated our planet to the extent that scientists have proposed a new geological period — the Anthropocene — where "A single species is in charge of the planet, altering its features almost at will."[1]

Has this unprecedentedly powerful influence by one species been for the better or worse? Taking the narrow economic view, it would appear the pluses definitely outweigh the minuses and we live better lives and are far more productive than our ancestors. In the social sphere too, despite wars, famines and epidemics we appear to have improved the quality of life for most humans.

---

[1] Pearce (2006).

It is in the environmental or biosphere that most damage has occurred from pollution of our air, water and land to depletion of resources.

Does this matter? If your concept of Sustainable Development is the Triple Bottom Line, then perhaps it is possible to justify the gains in the Economy and Society against Environmental losses and degradation. But it is increasingly clear that this is a dangerously flawed reading of our balance sheet.

We should, instead, regard the Economy, Society and Environment as three spheres or three envelopes of increasing size and dependency. The smallest, the Economy can only exist within the larger Social sphere which in turn is contained within the largest of the three, the Environment or biosphere. When populations and economic activity were small relative to the biosphere, it was possible to regard it as infinite or even disregard our impact on it.

However, growing scientific evidence including Climate Change indicated this is no longer possible and that humanity is today pushing our life-support systems to their limits. We should remind ourselves that despite astonishing technological advances we have not succeeded yet in recreating another biosphere to replace the natural one we are destroying.

In this context, the Haze is yet another example of this failure of vision and cannot be solved if we do not see it in the correct perspective. Instead it will only get worse not better since the wrong perspective will generate the wrong prescriptions.

Villain or victim?

Many other countries have suffered from the Resource Curse but Indonesia has been and still is particularly vulnerable. Her size and geographical diversity (she is the world's largest archipelago with 13,677 islands spanning some 5,000 kilometers) are matched by the size and cultural, ethnic, social and income diversity of her population. She is also rich in biodiversity and natural resources.

Even the adoption and implementation of the most sustainable development policies would have proved a huge challenge to any government. Unfortunately, her timing was bad.

Sustainable Development was still a concept of the future but even after the 1972 Stockholm Conference on the Human Environment and the

1992 Rio Summit, the Washington Consensus and market deregulation still prevailed and environmental damage and depletion considered an 'externality'.

Her downward spiral to unsustainable environmental practices was a conventional one. Her huge tracts of almost untouched tropical forests seemed a limitless natural resource. Logging concessions were given out by the government, but often with little regulatory oversight. Sustainable forest management practices were also frequently neglected in the pursuit of short-term profits. When valuable timber had been extracted, the companies moved to greener pastures and the logged forests were labeled as degraded. Some of these became eligible for commercial plantations mainly oil palm and pulp and paper or as transmigration sites, while other areas became vulnerable to illegal logging and further degradation.

Timber (forestry) was one of the pillars of President Suharto's New Order and the export of logs boomed in the 1970s making timber the most important non-oil export. This gradually became replaced by palm oil and pulp and paper as forests were degraded and depleted.

Three catastrophic events in 1997–1998 highlighted the economic, social and environmental challenges facing Indonesia on its journey toward sustainable development had become.

They were respectively:

- The Asian Financial Crisis of which Indonesia was one of the major sufferers.
- Social discontent and political unrest targeted at the 30-year regime of President Suharto and the inequitable and corrupt exploitation of the countries natural resources resulting in the President's resignation.
- The 1997–1998 land clearing fires and associated Haze.

Although, by then, the concept of Sustainable Development had been recognized at Stockholm and the Rio Summit '92 followed by the Santiago Consensus in 1998 that market failures need to be recognized and corrected and environmental factors taken into account and externalities internalized, these were still in a conceptual stage.

Salim (2005) observes: "For Indonesia, it was too late. We had already suffered. We had a disastrous financial crisis, soaring unemployment, a steep drop in the human development index, a high index of perceived corruption, accelerated deterioration of the environment, shrinking natural resources, increased illegal resource extraction and a serious weakening of governance. From 1998 to 2004, Indonesia's development was accompanied by great soul searching as we asked what had gone wrong and what we must do in the future."

"We now understand the need to reform the total development strategy of Indonesia and to achieve triple-track development — not only economic development but also social and environmental development. But how do we do this? What model can we use?"[2]

Many if not most countries have failed to implement these enlightened goals even at the best of times and it would have been a miracle had Indonesia achieved it at a time of political and economic crisis and in the face of conflicting advice.

She is still grappling with this problem through successive governments and while it is beyond the scope of this chapter to analyze her success or failure on the economic and socio-political front it seems fair to conclude that the pressure from these two spheres have resulted in relative neglect of her environmental problems which have continued their downward slide.

The unsurprising result is the 2015 Haze which was far more severe in scale and damage than the one in 1997–1998.

3 in 1 environmental problem

Haze

The haze is by no means the only or even the worst environmental disaster resulting from widespread fires; it is merely the most visible, tangible, dramatic and newsworthy. Despite this, it took several decades to clear the smokescreen of ignorance and deliberate misinformation on its cause. "Natural" forest fires were blamed although healthy tropical forests do not spontaneously combust unless cleared and degraded. Traditional shifting agriculture involving slash-and-burn techniques was also blamed although this had been practiced for millennia. As will be described in the following sections, the causes of fire and associated haze are for more complex.

---

[2] Salim (2005).

Finally, non-environmentalists and the media awoke to the fact that the haze was an escalating problem that had serious economic repercussions and was not 'just' environmental; it was man-made and linked directly to land clearing for palm oil, paper pulp and other export-linked plantations.[3] In short, it is a classic example of how market fundamentalism driven by the profit motive ignores social and environmental externalities.

The tipping point was reached in 1997 where a concatenation of decades of unsustainable logging and land clearing practices by fire for plantations, a huge population increase compounded by legal and illegal transmigration, and a particular severe El Niño resulted in an unprecedented conflagration.

For the first time, the Economic losses from the Haze were computed including direct damage and indirect short-term impacts such as human loss of food and income, accelerated soil erosion and sedimentation of water bodies, and disruption of hydrological and nutrient functions of forests. Acute health problems and disruptions and loss to business, tourism and transport were also computed.

The loss was estimated to be 4.5 billion US dollars although this was recognised as certainly an underestimation since long-term and indirect losses both to humans and the environment were impossible to capture. For example, there was a missing cohort of 15,000 children under three years of age which is attributed to air pollution from the 1997–1998 Haze.

Large though, this figure which is the estimated loss from the 2015 Haze is thought to be at least eight to 10 times greater.

Loss of forests

Difficult though it is to put an economic estimate to the Haze, this is nothing compared to the impossibility of calculating the cost of the ecological losses caused by land clearing fires, including the degradation and loss of forests, biodiversity and critical ecosystem services.

Percentages and averages do not begin to accurately measure the loss although these numbers are not small. As estimated 2.6 million hectares of land was burned during the 2015 fires.

---

[3]E.g. See Varkkey (2011).

Part of the challenge results from the far-reaching consequences both in space and time. For example, corals can be smothered by increased run-off from burnt and degraded land thousands of miles upstream resulting in damage to fish stock and other aquatic life.

While forests have the ability to regenerate, ecologists still do not know how much repeated damage a tropical rainforest can sustain before its ecosystem services and recovery mechanism totally fail. We may have reached the tipping point without even being aware of it. We also do not know how much forest cover loss our Earth can sustain, not just for the health of forests but of our Planet.[4]

Climate change

In spite of the warnings of climate scientists and environmentalists for over a quarter of a century, the levels of ignorance, smokescreens and lack of political will regarding climate change put that over the Haze in the shade.

Despite the Kyoto protocol, policymakers and businesses were reluctant to tackle a problem perceived as environmental and very much in the future; until the increasing frequency and severity of environmental and climate change related disasters forced a refocus.

One was the 1997 Haze which rocketed Indonesia to third place that year as a major global emitter of CO_2. It also forced the global community to realise that reducing emissions without a regulatory framework that addressed deforestation would be useless (Reducing Emissions from Deforestation and Forest Degradation (REDD) is not a part of the Kyoto Protocol).

The fact that Indonesia has the world's third largest extent of tropical rainforests as well as the majority of tropical peatlands which can release huge amount of greenhouse gases as well as haze from underground fires means that it is now a major focus for reducing emissions. If a workable mechanism can be found under REDD+, this could be our best hope of halting the juggernaut of deforestation, fires and haze.

---

[4]E.g. See Popkin (2017).

A destructive cycle

It should be evident that the three problems mentioned above, each complex and multifocal, are also closely interrelated. Loss of forest has contributed to the haze and climate change but a three-way reciprocity is also true in that accelerating climate change or more severe forest fires and haze will also exacerbate the other two problems.

A solution to be effective should therefore address all three problems.

It should also be evident that the problem is far larger and more complex than any nation can handle on her own. Indonesia has already been attempting to do much to address the problem including some innovative new measures that will be described later. However, multi-holder collaboration and support are also necessary.

The scale and severity of the 2015 Haze described as "certainly the greatest environmental disasters of the 21st century — so far[5]; compared to 1997 it strongly suggests the problem is escalating and can become worse in the future — unless a workable solution is found soon and put into effect.

Searching for sustainable solutions

Any workable solution for the Haze should take into consideration all the following points.

(1) Approach the three spheres (Economic, Social and Environmental) simultaneously and holistically; not selectively or in piecemeal fashion.
(2) Perceive that the haze is not an isolated environmental problem but inextricably linked with other environmental problems such as climate change, deforestation and biodiversity loss. Therefore no sustainable solution can be achieved by tackling the haze alone. Instead any working solution must recognize that deforestation and climate change will make the Haze worse and vice versa. On top of this, social and economic issues will inevitably be impacted by a deteriorating environment (see 1).

---

[5]Monbiot (2015).

(3) Resist quick fixes whether motivated by political expediency or narrow economic imperatives. Not only will they not solve the problem but are likely to make it worse as had happened before.

(4) Recognize that prevention is better than cure. Downstream solutions like fire-fighting are an expensive, reactive last resort, not the solution of first choice.

(5) Realize that no single country or agency can tackle this complex problem on its own. Multiple collaborations which are informed, integrated and sustained are required whether between government and government, government and people or people and people.

(6) Consider all stakeholders as potentially part of the solution. Policymakers whether local, national or regional, agroindustry and investors, smallholders, hired laborers, illegal transmigrants or overseas consumers must be engaged.

(7) Recognize that any solution is only as good as its implementation. Since the cooperation of those on the ground is critical, any amount of laws and regulations without local stakeholder input, motivation and cooperation is useless.

(8) Finally, acknowledge that this is a complex multifactorial problem with no easy solution. We should therefore seek for a diversity of strategies rather than a silver bullet.

Responding to the haze

Indonesia has taken some important steps towards addressing the interrelated problems of deforestation, land clearing fires and the resulting haze pollution. These include decisions by the Indonesian government to continue to enforce a 2011 moratorium on the clearing of primary forests and forested peatlands and to review the approval of new concessions on peatland. The government has also pledged to reduce national greenhouse gas emissions by 26 percent by 2020 and 29 percent by 2030 (BCSD Singapore *et al.*, 2016; SIIA, 2016). A National Peatlands Restoration Agency has been set up to lead the conservation and restoration of Indonesia's peatlands. Additionally, President Joko Widodo is reported to have said that he would ban all clearing and drainage of undisturbed peatlands from 1 June 2016 (SIIA, 2016).

At the regional level, responses have focused on a range of policy instruments, notably the ASEAN Agreement on Transboundary Haze Pollution, which has now been ratified by all the Association of Southeast Asian Nations (ASEAN) countries, and Singapore's Transboundary Haze Pollution Act of 2014 (SIIA, 2016). There has been growing emphasis on trying to identify and prosecute or at least 'name and shame' the corporations within the plantation sector on whose land fire is detected. But while a significant proportion of the fire hotspots detected in 2015 occurred within either pulpwood plantations (c. 20 percent) or oil palm concessions (c. 16 percent), prosecuting those responsible, including identifying the individuals who actually started the fires, is a far more difficult task (Greenpeace, 2015).

The underlying drivers of forest and other land clearing fires and transboundary haze are complex and dynamic. Environmental factors including the reduced area of natural forests and climatic phenomena such as El Niño can aggravate the number and intensity of fires, but the roots of the problem are ultimately social and economic. Thus, while there are no quick fixes or silver bullets, international and national responses to the haze problem will have limited impact without parallel and complementary measures at the local level, where the day-to-day decisions about land use are made and where the interplay of the socio-economic and political factors governing these decisions are continually evolving. Such measures should include greater and more systematic action to develop and implement fire *prevention* strategies and thus to strengthen capacity to detect, prevent and control fires early on particularly at the local level. This would likely be a more effective and efficient use of resources than the periodic major investments in fighting and controlling the spread of fire during severe haze episodes as in 1997–1998, 2013 and 2015.

The rest of this chapter considers the lessons learned by Burung Indonesia, an Indonesian conservation non-governmental organization (NGO), in relation to preventing and addressing land clearing fires in Sumatra based on their experience of implementing and managing Indonesia's first Ecosystem Restoration Concession (ERC): Hutan Harapan — the Rainforest of Hope, which extends over an area larger than Singapore. Hutan Harapan is a joint initiative of Burung Indonesia and two international partners, the Royal Society for the Protection of Birds

(RSPB, UK) and BirdLife International,[6] to restore and manage nearly 100,000 hectares of previously logged lowland tropical forest. Despite having been logged in the past, Harapan retains considerable forest biodiversity, which is a source of valuable ecosystem services and has great potential for restoration. However, Harapan, like many other parts of Sumatra and Kalimantan, continues to face the threat of further deforestation, driven mainly by the high demand for land for the production of agricultural commodities such as oil palm and natural rubber. This process often starts with illegal logging, followed by vegetation clearing and ultimately burning to remove all debris and prepare the land for planting. These land clearing fires can then spread to other forested areas.

Addressing the problems of illegal logging and land clearing fires is key to achieving Harapan's overarching objective of restoring the balance and productivity of the natural rainforest. Thus, efforts have been underway for several years to better understand the drivers of deforestation in Harapan and to develop and implement solutions that meet the short-term development needs and aspirations of local communities and in-migrants without compromising the objectives of Hutan Harapan and of the ERC policy. Resolving these issues effectively in the long term, including the problem of widespread fire and transboundary haze will require tackling the underlying economic drivers of deforestation in Sumatra and Kalimantan. This, however, requires concerted action by a much wider set of stakeholders, including local and national governments and the private sector. These issues are described and discussed further.

Indonesia's ecosystem restoration concession policy and the establishment of Hutan Harapan

Launched in 2004, ERCs marked a watershed in Indonesian forestry policy. Formally titled as the Ecosystem Restoration Timber Forest Utilization License for Natural Forest in Production Forest (IUPHHK-RE), the policy was a response to the threat of further degradation and deforestation of Indonesia's logged production forests to after the expiry of the logging concessions.

---

[6]BirdLife International is the world's largest nature conservation partnership widely recognized as the world leader in bird conservation. Burung Indonesia and RSPB are the national BirdLife Partner in Indonesia and the UK, respectively.

The ERCs offered a mechanism that could help to prevent the conversion of logged forests to monoculture plantations or non-forest uses by creating a new type of forestry concession that for the first time allowed degraded production forests to be managed primarily for restoration and sustainable non-timber-based uses. Under the policy, ERC holders are required to restore degraded and damaged forests to their 'biological equilibrium'. Additionally, as ERC licenses are issued for a period of 60 years and renewable for a further 35 years, they present a real opportunity for achieving significant results in terms of ecosystem restoration, biodiversity conservation, climate change mitigation and non-timber-based income generation thereby contributing to diversification of local economic opportunities and livelihoods (Silalahi *et al.*, 2017).

The first two ERCs to be approved under this new policy in 2008 and 2010, resulted in the establishment of Hutan Harapan.[7] Covering 98,555 hectares in Jambi and South Sumatra provinces, Harapan contains over 20 percent of Sumatra's remaining tropical lowland forest, one of the world's most threatened forest types; and, despite having been commercially logged in the past, Harapan also harbors a high proportion of the island's biodiversity, with over 1,300 recorded species, including over 130 globally threatened species and subspecies from tree frogs to Sumatran tigers and elephants (Silalahi *et al.*, 2015). However, Harapan is essentially a forest island in a sea of plantations — its immediate neighbors include one oil palm plantation and eight acacia plantations with a combined area of over 270,000 hectares, nearly three times the size of Harapan (Figure 1). Additionally, around 10,000 people live in and around Harapan, comprising a mixture of local communities as well as migrants from other parts of Sumatra and further afield. Harapan's boundaries are long and porous: the area is not fenced and being lowland forest, there are no natural barriers to humans in the form of steep slopes or wide fast-flowing rivers. Fencing the entire area or patrolling the concession boundaries would be too costly. Given the intense competition for land in Indonesia, not surprisingly there has been encroachment of the ERC, partly to establish oil

---

[7]Hutan Harapan was also the inspiration for BirdLife International's global forest program known as Forests of Hope. http://www.birdlife.org/worldwide/programme-additional-info/forests-hope

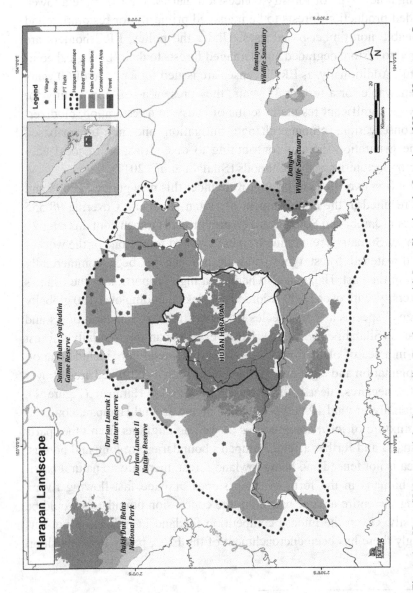

Figure 1. Hutan Harapan and the Surrounding Landscape, with Inset Showing the Location of Hutan Harapan on Sumatra

© Burung Indonesia

Source: Silalahi *et al.* (2017).

palm plantations, associated with an increasing incidence of fire, particularly between 2012–2015.

The significance of oil palm as a driver of land use change, economic development and haze pollution

In order to address the problem of deforestation and the associated land clearing fires and encroachment in Harapan — and in Sumatra and Kalimantan more generally — it is important to first understand the role of oil palm in driving land use change and economic development as well as its association with large-scale fires (Hadinaryanto, 2014a,b; Greenpeace, 2015). While not all deforestation is caused by oil palm expansion, this is known to be among the top three drivers of forest and land clearing in Indonesia, particularly of lowland tropical forest, and has been closely linked to the large-scale fires that lead to regional haze (Hadinaryanto 2014a; Abood *et al.*, 2015; Euler *et al.*, 2015; Greenpeace 2015). The traditional practice of slash and burn is frequently cited as the major proximate cause of such fires although this method of clearing land is widely practiced across Indonesia without causing regional haze. Indeed, the problem lies not so much with this method of clearing land, but rather in the purpose of doing so, namely to establish a plantation cash crop (i.e., oil palm) on a large enough scale to be profitable. The problem is further compounded when forest and land clearing by fire takes place on peat soils, which are also found mainly in Sumatra and Kalimantan.

There are many reasons why oil palm cultivation is attractive to so many actors. Palm oil production has been a major engine of economic development in Indonesia since the 1970s, particularly in rural areas. With some 7 million hectares of land under mature oil palm, mostly (c. 70 percent) in Sumatra and the remainder in Kalimantan (Indonesia Investments, 2016), Indonesia has become the world's largest palm oil producer, accounting for over 50 percent of global production by 2012 (Oxfam, 2014). Much of the expansion of oil palm has been at the expense of Indonesia's natural forests (Hadinaryanto 2014a,b; Greenpeace, 2015). Thus, between 2000–2012 Indonesia's annual deforestation rate was over 671,000 hectares, with more than 80 percent of the deforestation in this period occurring in Kalimantan and Sumatra (Government of Indonesia, 2014 in UNORCID,

2015). Sumatra lost two-thirds of its primary forest cover between 1990–2010 (Margono *et al.*, 2012), and Kalimantan, where oil palm expansion is a more recent development, lost nearly 5 million hectares of primary forests between 2000–2010 (Abood *et al.*, 2015). This has contributed to Indonesia having one of the highest rates of deforestation in the world (Margono *et al.*, 2014). It is also one of the world's largest producers of greenhouse gases with deforestation and destruction of peatlands accounting for 60 percent of total emissions on average in the 10 years up to 2015 (Climate Transparency, 2016; Government of Indonesia, 2016).

The ever rising global demand for palm oil has strengthened access to finance, government and corporate support and other incentives for its cultivation. Support for smallholder producers is often forthcoming because of the opportunities for intermediaries to extract benefit at different stages along the palm oil supply chain (e.g., see Purnomo *et al.* 2015). Lower labor inputs and better returns relative to many other agricultural crops, particularly for small holdings, also serve to make oil palm cultivation very attractive. However, the benefits of oil palm production are not evenly distributed, as not all communities, particularly indigenous people, are willing or able to shift to oil palm (Oxfam, 2014; Euler *et al.*, 2015). Even within the supply chain, smallholder producers rarely benefit as much as the many intermediaries further downstream. Thus, the expansion of oil palm has also resulted in widespread social conflicts, displacement of local communities and loss of customary lands, as well as affecting food security in some areas by displacing food crops (Hadinaryanto, 2014b; Oxfam, 2014). For example, in 2012 according to the Ministry of Agriculture, 59 percent of Indonesia's 1,000 palm oil companies were linked to land-related conflicts with local communities; conflicts were identified in 22 provinces and 143 districts, mostly across Kalimantan, but with a significant proportion also in Sumatra (Hadinaryanto, 2014a). Indeed, land-related conflicts between local communities and the plantation sector are a contributory factor in an unknown proportion of the fires that occur within the boundaries of many plantations.

Nonetheless, oil palm is likely continue to play a significant role in Indonesia's economy for the foreseeable future for several reasons. First, although Indonesia is now a middle-income country, nearly 40 percent of the population are still just above the national poverty line with higher levels of poverty in rural areas (IFAD, 2015). Second, the agriculture sector, which includes oil palm, still employs nearly 40 percent of the Indonesian labor force or some 50 million people (CIA World Factbook, 2016). Third, global demand for palm oil is projected to double by 2050. Finally, government policies continue to support oil palm expansion, including a target to increase annual exports from just over 25 million tons at present to 40 million tons by 2020 (Hadinaryanto 2014a; Oxfam 2014; Indonesia Investments 2016).

Meanwhile, the changing dynamics of palm oil production within the country presents new challenges. Smallholders account for some 40 percent of the total area under oil palm and a growing share of total palm oil production (Oxfam 2014; Euler *et al.*, 2015). Historically, smallholder production was associated with various government and/or private sector out grower schemes (e.g., see Euler *et al.*, 2015), but more and more smallholders have been striking out independently of such schemes, often with the support of diverse local actors, including politicians and financing institutions. Smallholder production is expected to increasingly dominate the supply chain as the regional economy is strongly driven by oil palm production and there are few equally or more attractive options for income generation. Promoting sustainable oil palm production in the landscape thus requires engagement of a much wider range of stakeholders than hitherto. While the direct influence of the major oil palm companies, including the international companies, has begun to wane, their role in creating a regional economy that is so heavily reliant on oil palm should not be forgotten. These companies therefore still have a major responsibility to promote greater sustainability of oil palm production, not just within their own plantation boundaries, but also within the wider landscape.

Additionally, there is need for a better understanding of the local palm oil supply chain, particularly the role of intermediaries between the producers and the mills, and the incentive structure that dictates the distribution of benefits from slash and burn for establishing oil palm (e.g., see Purnomo *et al.*, 2015). As Euler *et al.* (2015: p. 2) note: "Designing policies that can contribute to sustainable development requires good understanding of the factors that influence smallholder land use decisions in general, and their decision to adopt oil palm in particular."

Many parts of Indonesia are witnessing a 'land race' as different groups and stakeholders compete for agricultural land with multiple claims often being laid to the same piece of land (Silalahi *et al.*, 2015; Silalahi *et al.*, 2017). In-migration and land use change in Harapan, as elsewhere in Sumatra, Kalimantan and further afield, are ultimately driven by economic imperatives, sometimes aided and abetted by politically motivated interests who may support land grabs in exchange for votes or to expand palm oil production controlled by outside interests (Hadinaryanto, 2014b).

Finally, given the Indonesian government's target to significantly increase oil palm exports by 2020, strategies and investment are needed to enhance smallholder productivity to ensure the increase in overall national production results from improved productivity rather than further expansion, particularly by clearing natural forests. The productivity of Indonesian oil palm is generally lower than that of Malaysia, while within Indonesia, smallholder productivity is reported to be 35 percent below the average of private companies and 45 percent below government companies. In practice, smallholders' yields vary considerably, but differences can be marked, with some producing as little as 1–3-ton Crude Palm Oil (CPO) per hectare compared to production of over 6-ton CPO per hectare by large oil palm plantations (Oxfam, 2014).

Management responses to forest and land clearing fires in Hutan Harapan

An in-depth understanding of the local socio-economic and political context has shaped Hutan Harapan's overall management strategy, including the response to the problem of fires and encroachment in the ERC. This includes a good understanding of the role of oil palm and other agroindustrial commodities in driving land use and other socio-economic changes in the wider landscape. Harapan has taken a multi-pronged approach to address these issues. At a practical site-based level, Harapan has invested heavily in fire prevention and control. The approach, which for the most part has been implemented successfully, involves a combination of adequate infrastructure, well-trained manpower and strict fire monitoring, prevention and management protocols. A Forest Protection Patrol team undertakes daily and monthly patrols, covering different areas by foot and vehicle (i.e., motorbikes and cars). In 2013, a daily Fire Danger Rating System was initiated; this includes monitoring and analysis of weather and vegetative conditions to identify (a) the potential for a fire to ignite and spread and (b) the likely need for prompt fire control. Harapan has also invested in firefighting equipment and a well-trained-fire-fighting team. Thus, in 2012 and 2013, there were only seven days with fire; and in 2013, only 20 hectares were burned — a remarkable achievement given the extent of damage in areas outside the ERC. However, the frequency and intensity of the fires in 2015, exacerbated by the El Niño conditions, were such that even Harapan experienced greater damage than before, with an estimated 13,400 hectares affected by fire (PT REKI, 2016). While, the full impacts of the fires are still being assessed, in some areas the forests are already starting to recover.

Additionally, since its establishment, Hutan Harapan has been working with a range of stakeholders, including local and national governments, to identify appropriate strategies to prevent further in-migration into Harapan; engage constructively with the existing population living within the ERC; and prevent further natural habitat loss. Altogether, 20,337 hectares (or 21 percent) of Hutan Harapan have been cleared illegally since 2005 (i.e., even before the ERCs were granted), and a little over 2,000 families now live within the concession boundaries. These include in-migrants, who have planted oil palm over some 50 percent of the cleared area and smaller numbers of indigenous communities such as the Batin Sembilan, who did not participate in the illegal land clearing and who generally do not undertake oil palm cultivation. Although fire was the method of choice for clearing land, there has been relatively little loss of Harapan's more intact natural forest in the interior of the concession as the first areas to be encroached were in the more degraded and accessible parts of the concession close to its boundaries.

Hutan Harapan's approach to resolving land use conflicts in Harapan involves working with all relevant stakeholders across the wider Harapan landscape including local and national government, the plantation sector, other civil society organizations and local communities. Identifying mutually acceptable solutions to these complex and deeply entrenched economic, social and political problems is a long process that requires dialogue and negotiations with numerous stakeholders and considerable investment of time, effort and patience (Silalahi and Erwin, 2014). Disentangling the various land claims within the concession is complicated by the many interest groups involved and the large number of overlapping claims. These groups include: indigenous peoples, migrants from other parts of Indonesia, land speculators, local elites and agrarian reform activists, each group with its own agenda, creating conflicts with each other as well as with Hutan Harapan's management.[8]

Finally, a physical 'space' where conflict resolution and mutually acceptable solutions could be developed and implemented has also been created and is reflected in Harapan's management zonation. The latter includes a Collaboration Zone (which combines forest and land rehabilitation with livelihood development) and a Livelihood Zone (which focuses on creating livelihoods through agroforestry development), covering 21 percent and 5 percent, respectively of the total ERC area. These two zones are intended to encompass the economic activities of Harapan's resident population as brokered through various agreements with Hutan Harapan's management and the Ministry of Environment and Forestry, which by law must be party to any land use agreements with communities in Harapan.

To date, seven agreements have been signed with communities, including four with indigenous Batin Sembilan communities covering up to 104 families who were already present in the ERC at the time of establishment and one with a local Malay community covering 13 families. The other two agreements are with in-migrant communities covering around 200 families. Clearing of land for planting oil palm is generally by in-migrants. Where this has been done and the oil palm is ready for harvesting, the agreement allows the communities to retain the plantations and to harvest oil palm for one full life cycle (c. 25 years). Additionally, communities may plant agreed agroforestry crops within the 'Collaboration Zone'. Hutan Harapan will work with the communities to agree on transitioning all existing oil palm to other agroforestry crops at the end of one cycle. Potential agroforestry species include rubber and a variety of fruit trees and timber species.

The most challenging negotiations to date have been with the settlements to the east of Harapan — which are located close to oil palm mills

---

[8]E.g., see Purnomo *et al.* (2015) analysis of the dynamics and distribution of benefits between different actors involved in land clearing.

and thus already part of a well-established palm oil supply chain.[9] Two pre-agreements have been signed with these settlements, which include around 1,000 families. However, identifying options that will outweigh the benefits of growing oil palm will be more challenging in these two settlements. A key future activity will be to work with other stakeholders to encourage the mills to accept only legally sourced oil palm. A further angle that remains to be tackled is the role of local institutions that often finance the expansion of smallholder oil palm. Finally, local government plays a pivotal role in the expansion of oil palm, as it controls the approval of new agricultural concessions and whether smallholders can access finance and obtain support for clearing land to establish near areas of oil palm. Therefore, strengthening local government capacity to support natural resource governance and landscape-level planning that goes beyond spatial planning and narrow administrative boundaries is also essential to promote wider landscape-level sustainability and address the drivers of forest encroachment, land-clearing fires and other illegal activities in Harapan and elsewhere in Sumatra and Kalimantan.

Addressing existing land use conflicts in Harapan and pre-empting fresh ones will continue to be a major focus of work. In relation to oil palm, Harapan's objectives are to: (1) partner with government and the palm oil industry, local communities and other relevant actors to monitor the supply chain and ensure there is no further expansion of oil palm within Harapan; and (2) promote sustainably produced palm oil in the wider landscape including preventing the further clearance of natural forest for this purpose. However, addressing the underlying drivers of land use change and developing sustainable long-term community partnerships cannot be the sole responsibility of ERC-holders. Sustained long-term investment and multisectoral, multistakeholder collaboration are needed to identify and implement workable solutions along with a paradigm shift in local government planning.

Stabilizing forest loss, rationalizing land use change and promoting sustainable landscapes

Addressing land use conflicts and preventing further deforestation, land clearing fires and the associated haze pollution are clearly in Indonesia's

---

[9]Proximity to mills is key for oil palm production as fresh fruit bunches must be processed within 48 hours of harvesting.

national interests given the enormous costs to both Indonesia's own citizens as well as to its regional neighbours (SIIA, 2016). Indonesia's natural forests are also key to meeting national objectives and global commitments on climate change, biodiversity and land degradation. Indonesia vies with Brazil as the world's most biodiverse nation, yet with only one fourth of Brazil's forest area, Indonesia's deforestation rates have exceeded those of Brazil in recent times.[10] Her forests, particularly those on carbon-rich peat soils in Kalimantan, Sumatra and Papua are also considered key to achieving global targets on keeping global warming below dangerous levels. An estimated 2.6 million hectares of land were burned in these areas during the 2015 fires in Indonesia.

From a biodiversity and climate change perspective, the on-going loss of Indonesia's natural forests is a major ecological tragedy. With an estimated 20 million people living in and around Indonesia's forests, it is also often a social tragedy, most immediately for forest-dependent people. Ultimately, the continued loss of forests will have profound impact on even the non-forest dwellers through the degradation and loss of ecosystem services that are critical for human well-being and continued economic development. In the short term, periodic bouts of severe haze serve as an acute reminder of the consequences of environmental mismanagement. In the long term, natural forest loss will contribute to the problems of on-going climate change.

Indonesia's pursuit of economic growth at the expense of her natural capital and potentially her long-term sustainability is not unique — it is the common story of most of the world's economies. Balancing the competing priorities of promoting rural development, economic growth, public health, environmental stability and good relations with her ASEAN neighbors is a major challenge for the Indonesian government as it is for many other countries, especially those that still have significant levels of rural poverty. However, it is still possible to alter this trajectory and there are promising signs that there is growing political will to do so. Some of the key steps taken by the Government of Indonesia to address the problem of fires and haze pollution have already been mentioned. The government's five-year medium term development plan for

---

[10] Margono *et al.* (2014).

2015–2019 identifies land use conflict resolution and natural resouce governance, including clarifying tenure over land and other natural resources, as clear priorities for action. A new Directorate of Conflict Resolution, Tenure and Customary Forests (*Hutan Adat*) was established in 2015 and has begun working with ERC licence-holders under the Ministry of Environment and Forestry to resolve land-related conflicts. The Directorate is using Hutan Harapan as one of a few national pilots for consolidating experience on managing land use conflicts to guide the development of national regulations related to conflict resolution. Since September 2015, this directorate has taken the lead in negotiating with local communities and in-migrants living in and around Harapan, together with Burung Indonesia, relevant national, provincial and local stakeholders as well as human rights NGOs to find mutually acceptable solutions.

While Harapan represents a drop in the ocean in terms of its contribution to the overall haze problem, it represents a microcosm of what is happening to land generally in Indonesia and especially in Sumatra and Kalimantan. A number of lessons have emerged from the experience of managing Harapan and dealing with the pressures of encroachment and palm oil expansion which are linked to drivers that emanate outside Harapan and also have impacts well beyond Harapan, notably in the form of regional haze. In the short term, investment in capacity to prevent and control fires is needed. However, medium- to long-term strategies must be implemented alongside to address the underlying drivers of forest and land-clearing fires in Sumatra and Kalimantan.

Addressing these drivers is the collective responsibility of many stakeholders within the wider landscape and further a field who must work collaboratively to address the underlying drivers of forest loss and unsustainable land use change in Sumatra and Kalimantan. These stakeholders include the plantation sector as well as national and local government, while the drivers relate to the continuing expansion of the agro-industrial sector, notably oil palm, but also paper and pulp, as well as the activities of other sectors such as mining and infrastructure development. A new approach is needed whereby plantation companies and local government begin to look beyond their immediate areas of jurisdiction — their concession or district and provincial boundaries, respectively — to consider the social, economic and environmental sustainability of the wider landscape, i.e., the bigger long-term picture.

There is much that the plantation sector can do immediately to promote greater sustainability in the wider landscape by strengthening traceability of the oil palm coming into their supply chain, investing in smallholder productivity, investing in fire prevention beyond their borders, and supporting viable economic alternatives to oil palm (e.g., agroforestry crops) through corporate social responsibility programs and other sources of investment. The private sector could be further incentivized to invest more in fire prevention and non-oil-palm-based local economic development by being held to account for the air pollution caused by fires in their concessions through strengthened government laws and enforcement.

Bringing about a paradigm shift in local government planning requires targeted capacity development, however. At present, local governments are ill equipped to deal with issues relating to landscape-level analysis and planning, including matters relating to ERCs. In particular, there is a need to ensure that environmental values such as the maintenance of ecosystem goods and services, which are clearly recognized in national government policy, are also integrated into local government planning. Training and additional resources are needed to bring about a more holistic approach to planning, one that goes beyond narrow administrative boundaries and the provision of basic infrastructure and services (i.e., physical and spatial planning) to one that that seeks to balance multiple broader objectives at a larger scale such as economic development, biodiversity conservation, maintenance of hydrological cycles and reducing greenhouse gas emission. A better understanding of the drivers and consequences of unsustainable land use change, in particular further natural forest loss, is also needed. A detailed understanding of the palm oil supply chain together with official maps of land tenure and forest cover that show concession boundaries is key to ensuring transparency and accountability in relation to illegal forest clearance (Greenpeace, 2015).

Developing viable economic alternatives to oil palm is a major challenge that requires both investment and innovation, which in turn will only happen if there is genuine commitment to long-term sustainability by government, private sector and other donors. There are several potentially marketable forest products and services including: carbon, agroforestry products, Non-Timber Forest Products (NTFPs) from natural forests, and payment for ecosystem services (PES) such as ecotourism and energy and water supply. However, markets for many of these either do not exist as yet, or the financial returns at present are too low when compared to oil palm. There are many barriers to developing new markets or expanding existing ones through value addition and improved marketing. These include the high upfront investment needed to cover startup costs, as well as other financing, technical and capacity-related obstacles, and the reluctance of potential financiers and investors to take on the risks of untested business models. However, identifying such alternatives including forestbased sources of livelihood is key to preventing further deforestation and promoting restoration of degraded lands and forests such as Harapan. There is pressing need for mechanisms that reward ERCs for contributing to Indonesia's Nationally Determined Contribution to reducing global greenhouse gas emission and incentivize investment in ERCs. The operationalization of Indonesia's national REDD+ strategy published in 2012 along with the further implementation of the Paris Agreement on Climate Change could greatly strengthen investment and action in this area.

Indonesia's ERC policy was a bold step towards reversing the country's deforestation trend, while also supporting the restoration of forest functions and services. To date, over 500,000 hectares of degraded production forests in Sumatra and Kalimantan have been allocated as ERCs and a further 1.6 million hectares of degraded production forests have been earmarked for ERC allocation reflecting the Government of Indonesia's commitment towards forest restoration (Renstra KLHK, 2015–2019). Hutan Harapan's vision as an ERC is be an integral part of a wider productive and resilient socio-ecological landscape that can continue to sustain human wellbeing and development over the long term. However, neither this vision nor the enormous potential of Indonesia's ERC policy will be realized without additional policy measures, strengthened enforcement of existing laws and multistakeholder collaboration and action to address the drivers of land degradation and forest loss and promote the implementation of business models that support sustainable land use and forest conservation.

References

Abood, S. A., Lee, J. S. H., Burivalova, Z., Garcia-Ulloa, J. and Koh, L. P. (2015). Relative contributions of the logging, fiber, oil palm, and mining industries to forest loss in Indonesia, *Conservation Letters*, 8(1): pp. 58–67.

BCSD Singapore, PBE, IBCSD and VBCSD. (2016). Efficient agriculture, stronger economies in ASEAN: Private sector perspectives for policy makers, White Paper, Business Council for Sustainable Development Singapore, Philippines Business for the Environment, Business Council for Sustainable Development Indonesia and Vietnam Business Council for Sustainable Development.

CIA World Factbook. (2016). The World Factbook. Retrieved https://www.cia.gov/library/publications/the-world-factbook/geos/id.html. Accessed 17 May 2016.

Climate Transparency. (2016). Brown to green: G20 transition to a low carbon economy: Indonesia. http://www.climate-transparency.org/wp-content/uploads/2016/08/Indonesia-2016.pdf. Accessed 21 August 2017.

Euler, M., Schwarze, S., Siregar, H. and M. Qaim. (2015). Oil palm expansion among smallholder farmers in Sumatra, Indonesia, EFForTS Discussion Paper Series 8, GOEDOC, Georg August University, Gottingen.

Government of Indonesia (2014) in UNORCID (2015) (p.28). National forest reference emission level for deforestation and forest degradation in the context of the activities referred to Decision 1/CP.16, Paragraph 70 (REDD+) under the UNFCCC. Submission by Indonesia.

Government of Indonesia. (2015). Indonesia land degradation neutrality national report. Jakarta.

Government of Indonesia. 2016. First Nationally Determined Contribution: Republic of Indonesia. http://www4.unfccc.int/ndcregistry/PublishedDocuments/Indonesia%20First/First%20NDC%20Indonesia_submitted%20to%20UNFCCC%20Set_November%20%202016.pdf. Accessed 21 August 2018.

Greenpeace. (2015). *Indonesia's Forests: Under Fire*, Greenpeace International: Amsterdam.

Hadinaryanto S. E. (2014a, April 24). Special report: Palm oil, politics and land use in Indonesian Borneo (Part I). *Mongabay*. Retrieved https://news.mongabay.com/2014/04/special-report-palm-oil-politics-and-land-use-in-indonesian-borneo-part-i/. Accessed 19 May 2016.

Hadinaryanto S. E. (2014b, April 26). Special report: Palm oil, politics and land use in Sumatra (Part II). *Mongabay*. Retrieved https://news.mongabay.com/2014/04/special-report-palm-oil-politics-and-land-use-in-sumatra-part-ii/. Accessed 19 May 2016.

IFAD, (2015). Investing in rural people in Indonesia. International Fund for Agricultural Development: Rome, March 2015.

Indonesia Investments. (2016). Palm oil in Indonesia — Indonesian CPO production & export. Retrieved http://www.indonesia-investments.com/business/commodities/palm-oil/item166. Accessed 19 May 2016.

Margono, B. A. *et al.* (2012). Mapping and monitoring deforestation and forest degradation in Sumatra (Indonesia) using Landsat time series data sets from 1990 to 2010, *Environmental Research Letters*, 7(3).

Margono, B. A., Potapov, P., Turubanova, S., Stolle, F. and Hansen, M. C. (2014). Primary forest cover loss in Indonesia over 2000–2012. Retrieved http://www.nature.com/nclimate/journal/v4/n8/full/nclimate2277.html

Monbiot, G. (2015, October 30). Indonesia is burning: So why is the world looking away? *The Guardian.*

Oxfam. (2014). Fair company — Community partnerships in palm oil development, Oxfam Discussion Papers, Oxfam International, May 2014.

Pearce, F. (2006). *The Last Generation: How Will Nature Take Her Revenge for Climate Change*, Eden Project Books: London.

Popkin, G. (2017). Tropical forests may be carbon sources, not sinks. *Nature News & Comment.* doi:10.1038/nature.2017.22692. Accessed November 2017.

PT REKI. (2016). *A Transect Analysis of the Forest Fires in Hutan Harapan*, Bogor: PT Restorasi Ekosistem.

Purnomo, H. (2014). Haze crisis and landscape approach, The Jakarta Foreign Correspondents Club: Jakarta, 2 July.

Purnomo, H., Gaveau, D.. Carmenta, R., Shantiko, B. and B. Kristanty. (2015). Political economy of fire and haze, Powerpoint of presentation made at British Embassy, Jakarta, 26 October 2015.

Renstra KLHK. (2015–2019).

Salim, E. (2005). Looking back to move forward, in Resosudarmo, B. P. (ed.), *The Politics and Economics of Indonesia's Natural Resources*, Singapore: ISEAS–Yusof Ishak Institute, pp. xxi-xxiv. 2005.

SIIA. (2016). Southeast Asia's burning issue: From the 2015 Haze Crisis to a More Robust System., Policy Brief, Singapore Institute of International Affairs.

Silalahi M. and Erwin D. (2015). Collaborative conflict management on Ecosystem Restoration Concession: Lessons learnt from Harapan Rainforest Jambi-South Sumatra-Indonesia. *Forest Res* 4: p. 134. doi:10.4172/21689776.1000134.

Silalahi, M., Walsh, T., Utomo, A. B., Barnard, J. and Woodfield, E. (2015). Hutan Harapan and Ecosystem Restoration Concessions as a new approach to

Sustainable Forest Management in Indonesia, Voluntary Paper accepted by the XIVth World Forestry Congress, Durban, September 2015.

Silalahi, M., Utomo, A. B., Walsh, T. A., Ayat, A., Andriansyah and Bashir, S. (2017). Indonesia's Ecosystem Restoration Concessions. *Unasylva* 249. Vol. 68 2017/1. pp. 63–70.

UNORCID 2015. Forest Ecosystem Valuation Study: Indonesia. United Nations Office for REDD Coordination in Indonesia.

Varkkey, H. M. (2011). Plantation land management, fires and haze in Southeast Asia. *Malaysian Journal of Environmental Management* 12(2) (2011): pp. 33–41

World Bank (2016). The Cost of Fire: An Economic Analysis of Indonesia's 2015 Fire Crisis. Indonesia sustainable landscapes knowledge; note no. 1. Washington, D.C.: World Bank Group. http://documents.worldbank.org/curated/en/776101467990969768/The-cost-of-fire-an-economic-analysis-of-Indonesia-s-2015-fire-crisis. Accessed November 2017.

8 Innovation and Transboundary Haze

Alwi Hafiz

Sustainability Advisor, Golden Veroleum Liberia
alwi_hafiz@yahoo.com

A plausible scenario

Thermal robotic eyes mounted on a tall pole, situated among the rows of palm trees, spot a possible small fire 5 kilometers away. An alert is transmitted to the control centre, with Global Positioning System (GPS) coordinates of the hot spot. Maps are automatically retrieved on a screen. The image on the screen superimposes hotspot and boundary information on images of the terrain. The operator activates the Standard Operating Procedure on fire. The fire management team is alerted. Locations of the suspected fire and closest water resources are transmitted to the team. The information includes which water sources are actually usable, using information provided by embedded water level sensors which had been transmitted routinely earlier in the day. First responders move towards the suspect site. A drone is launched upon approach. It confirms and pinpoints the location of the fire — fortunately still small with an 8-meter radius, but the visual images of the immediate surroundings show some smaller fires not detected earlier. The images help the team leader strategize the fire control approach. The drone moves away from the fire towards known water sources, and visually confirms the water level in a nearby canal is usable. The fire management team deploy their equipment and put out the fires.

Hardly cutting-edge, such technology to detect and prevent inevitable fires from getting out of control and spreading is already being deployed.

However, the challenges of doing so are not trivial, given the scale and landscape of commercial plantations, community farms and forests in Indonesia. Cost is obviously a major consideration. Equally daunting are training and education, the sheer logistics and planning for rolling out such systems, and building and maintaining the necessary support and management infrastructure to keep the systems effective. And this needs to be replicated at a massive scale to reduce the likelihood of major and infectious fires and the resultant transboundary haze.

There are many other technologies being used, tested or developed. Cameras — whether optical or thermal — require some level of line-of-sight to detect possible fires. This may not be practical in all cases. The use of environmental sensors represents an interesting alternative. These devices sense physical parameters such as temperature, pressure and humidity, as well as chemical parameters such as carbon monoxide, carbon dioxide, and nitrogen dioxide. The sensors are linked in a network and exchange information, which is analyzed against calibrated baseline information using sophisticated algorithms to determine the presence and possible locations of fires within the area covered. Already being used in Portugal, these systems are relatively new and would require more field trials in Indonesia and probably further development before they can be deployed on a larger scale with confidence. Just as in the case of the thermal cameras, the sensor network is only one part of the systems. Communication links are required for information exchange between sensors, the sensor network and control centers. An operations infrastructure ensures that timely and effective actions can be taken when fires are detected.

It is, however, hard to imagine the widespread and comprehensive deployment of such tools and methods over very large and uninhabited or sparsely inhabited areas such as those which exist in Indonesia. The use of surveillance aircraft, equipped with thermal imaging and other sensing equipment remains useful in such cases, and provides a faster way of detection compared to satellite sensing and imaging, which still has a lag time and may not be acceptable in the management of forest fires. Fire-fighting aircraft are also indispensable where fires occur in remote areas which are difficult to reach by any other means.

A well-planned, connected, coordinated and continuously evolving system incorporating existing and developing technologies deployed and managed by central and local government agencies, private companies and even community groups is plausible but quite a distant reality at the moment.

Innovation in prevention

Besides detection, prevention, especially in the form of tackling the root causes of the occurrence and thereafter spread of such fires, remains critical. The Indonesian government has imposed and pledged to enforce a moratorium on new oil palm concessions. It intends to exercise more control and enforce restrictions on deforestation and planting on peatlands. It has also set up the Peat Restoration Agency with the initial goal of restoring 2 million hectares of peatlands by 2020. Many of the largest plantation companies and all those which are members of the Roundtable of Sustainable Palm Oil (RSPO) have also long pledged zero burning and zero tolerance for burning within their companies and most are now focusing on ensuring the same policies get implemented and enforced into their supply chains. However, it is extremely challenging to impose any form of control over the actions of many thousands of smaller companies and indeed individuals within communities. For these groups, burning remains not only the cheapest and most effective form of land clearing, but sometimes the only one available. This is perhaps where innovation would be most impactful. In this context, several companies have begun to explore and implement interesting ways of involving the communities proximate to or within their developments in the effort to prevent forest fires.

Who should take the lead?

Large private companies produce roughly half of Indonesia's palm oil output, smallholders about one third and state enterprises the rest. Besides government, expectations naturally are on the larger plantation companies, often considered the villains in this unfortunate state of affairs, to lead the way. In reality, large commercial plantation companies, especially those which have large areas already planted, have much more to lose from fires within and close to their properties. Besides the loss from destruction of planted trees, palm oil companies suffer significant potential reduction of yields from prolonged heavy haze. Blockage of sunlight due to haze affects photosynthesis and stunts the growth of palm oil fruits. Heavy haze also affects insects which act as pollinators for oil palms. While studies are not conclusive as there are many other factors which affect yield, the industry experience is that prolonged haze of over a month can affect yields by more than 10 percent, a significant hit to profitability.

Burning to clear patches of land also does not work well for perennial crops like oil palm. Slash-and-burn methods typically improve soil fertility for a very short period but have a negative longer-term impact leading to the requirement for increased fertilization and treatment over the duration of the oil palm crop cycle. Hence the method currently practiced by most commercial oil palm growers consists of mechanical clearance followed by cutting and stacking the wood and plant debris, thereafter planting legume cover crops to aid decomposition. The returning of plant nutrients and organic matter into the soil in this way enhances soil fertility in a more sustainable and beneficial manner compared to burning.

It is not surprising then that besides having no-burning policies, many large companies already have in place and continue to deploy early fire detection and management systems. Besides short-term financial and business interest — possibly the most direct motivator — companies increasingly realise that sustainable environmental and social practices are important contributors to longer-term business performance and viability. Furthermore, palm oil companies which are members of the RSPO are mandated by its Principles and Criteria to have such measures in place. RSPO-certified mills and their supply bases are audited yearly to ensure that this is the case.

Consider again however the scale of the challenge. Taking the oil palm industry in Indonesia as an illustration, about 11 million hectares is presently under cultivation, of which about one third are owned or managed by small and medium enterprises and communities. Using one of the above technologies — the thermal robotic eyes — as an example, the costs are around 20 US dollars per hectare in initial capital expenditure and 10 US dollars per hectare per year just for equipment operation alone. Add to this supporting infrastructure and trained staff required to operate the system, the cost and effort required are not insignificant and would take many years to implement. While such technology could be within the reach of larger companies, it is unlikely that smaller companies and small-holder farmers would be able to afford them.

Boundaries of responsibility?

Even as the larger companies are rolling out fire detection and management systems within their concessions, the risk of fires spreading from neighboring areas occupied by smallholders, community farms and other areas into company plantations remains high if these areas are not covered by similar systems. The larger companies are therefore compelled, by a combination of self-interest and corporate social responsibility, to consider extending their monitoring and management capabilities to these neighboring areas.

This takes place within the context that fire has long been used to clear small patches of forest, farms or fields to clear land for farming, replanting crops and other uses. This happens not just in Indonesia but also in many other places around the world. As traditional community or family farming evolves and shifting agriculture declines, increasingly larger areas are now cleared for more commercially-oriented farms, planting a variety of crops including cash crops. The use of fire in the latter case is naturally more difficult to control. However, the alternative of mechanized clearing may not be affordable or even available to such farmers.

Innovative approaches

Some large companies — mainly palm oil and pulp producers — have responded by launching interesting and innovative projects not only to educate communities around their plantations but also to recognise and help meet their needs. Besides providing tools and training related to preventing and stopping fires early, these schemes help communities find alternatives to fire use, explore alternative livelihoods and even assist in land use planning.

In December 2015, Asia Pulp and Paper (APP) launched the Desa Makmur Peduli Api (Fire-aware, prosperous village) project. This project, which in the first stage will cover 500 villages — in itself illustrating the scale of the problem — aims to "support the economic development of 500 villages in the landscapes surrounding APP's supply chain. The aim of the programme is to demonstrate that economic development can be pursued in a sustainable way that supports rather than undermines the protection of Indonesia's forests."

Essentially, APP's approach is to pilot a series of community agroforestry programs. The APP would, together with the selected communities, take a landscape-based approach in deciding appropriate activities and assist these communities through the sharing of farming techniques and business skills to help the communities start and successfully operate agricultural enterprises. This could include rearing livestock and sustainable fruit and vegetable farming. This aims to enable alternative livelihoods that do not require the clearance of natural forest for further economic development. The APP believes that the programs will also help reduce instances of conflict over land by providing less land-intensive development options and prevent land encroachment and slash-and-burn activities.

Sinar Mas Agro (PT SMART), which is controlled by Singapore-listed Golden Agri Resources, is rolling out a similar but smaller project — the Desa Siaga Api (Fire-ready Villages). The pilot project involves eight villages in West Kalimantan and nine in Jambi, Sumatra. Through Desa Siaga Api, villagers (15 in each village) are trained and provided with facilities and equipment for extinguishing fires within their target areas. The volunteers are trained and supported to monitor hot spots and convey information in a timely manner to company-run or assisted field emergency response team through e-mail, text message, telephone, or other means of communication. Monitoring of areas and hot spots is also conducted by drones as well as satellite-based hot spot monitoring system whose data output will be forwarded to the Desa Siaga Api Task Force Center in the respective regions. Interestingly, there is an incentive element — villages which are successful in reducing or eliminating fires will be eligible for additional support in the form of social infrastructure development aid or technical assistance from PT SMART Tbk.

Besides fire management, the Desa Siaga Api program also trains and assists villages in alternative solutions to burning, including training via demonstration plots on land agreed upon by the residents, use of eco-friendly fertilizer, improvement of water systems and other related initiatives. Using a technique known as Participatory Mapping, communities will be trained and encouraged to map village boundaries and plan how land is used in their area. These maps will also help determine roles and responsibilities for fire prevention both within and between villages.

Solutions needed, not initiatives?

Critics may argue that these innovative initiatives are long overdue and that the resources and attention being channelled into the problem by the larger companies are inadequate. This is understandable, given the perception that these companies have contributed to the problem and profited enormously from their activities which, at least in the past, did not pay too much attention to environmental and social consequences. It also remains to be seen whether such projects would be truly effective and sustainable. Such company-sponsored projects are necessarily bound by time and budget considerations. The Desa Siaga Api project for example has been committed to run for three years. What happens next is not clear. It is difficult to imagine that the villages would be able to be self-sufficient or that local governments could step up to fill up whatever gaps there may be should companies decide not to continue. Presumably the companies will evaluate the projects in terms of effectiveness, cost and benefit before decisions are made.

It is therefore important that alternative longer-term scenarios are formulated before these initiatives expire and preparations are made in advance. One possibility, which could be really impactful, is that the Indonesian government introduces measures — legislative or incentive-driven — which would mandate or strongly motivate all relevant companies to take similar measures in and around their concessions, after learning from the various projects or pilots taking place.

While there are already many initiatives — driven by government, private enterprises as well as civic or community groups, it is not evident that there is an overarching, long-term plan to organise and bring together these initiatives to resolve the problem in a sustainable manner. Perhaps the collective impact of different parties doing different things will somehow make a significant difference and lead to desirable outcomes. But is there a better way?

Does the nature and scale of the problem lend itself to a more analytical approach using available science and data as starting points? Or is the knowledge already there, and the problem is primarily operational, dealing with issues of scale and dependence on just too many variables and uncertainties? In either case, since the problem is current and recurrent, various stakeholders will continue to decide what can and needs to be done and plan how to most optimally deploy available resources, technologies and innovative methods. However, subjecting disparate programs to proper planning, funding, coordination, management, execution and measurement remains a challenge. Perhaps the best that can happen — at least for now — is for those who can to do what they can. Hopefully there will be convergence at some point or a more coherent plan will somehow be forged along the way.

What can we do?

The problem of scale remains potentially overwhelming even if the larger companies and the Indonesian government ramp up their efforts significantly using available technologies and innovative methods to tackle both the root causes of the problem as well as the management of fires when they occur.

Could Singapore play a bigger part in this? That we have very little control over what happens in the lands and forests of Indonesia and Malaysia but suffer the consequences of ecological neglect or mishaps naturally translates into a sense of frustration.

Unfortunately, this sometimes leads to unproductive exchanges between our countries and peoples. An Indonesian government official said during the period of the haze of 2015 that "It's only been a week of smoke but people are already making so much noise. What about the oxygen that (Indonesia) supplies to them during the rest of the year?" Perhaps an understanding of the sentiment behind the statement would be helpful. The fact is that Singapore is a city-state and our geographic 'hinterland' are neighboring sovereign states. This hinterland provides not only economic but also important ecological services. Unfortunately, the nature of the latter is such that it is generally taken for granted, not assigned a tangible or measureable value and becomes apparent only when things go wrong.

How far will Singapore or Singaporeans be allowed to directly be involved in a complex matter concerning the government, businesses (some of which are listed in Singapore) and communities of a neighboring country?

In May 2016, the Indonesian Environment and Forestry Minister announced that Indonesia would terminate some current collaborations on haze and environment issues with Singapore while upcoming ones would undergo a "substantial review process." The review process would not be jointly done with Singapore and Singapore would (only) be informed of the outcome. This further illustrates that while government-to-government initiatives are perhaps more likely to have the financial muscle and reach to make a significant impact, the complexities of navigating Indonesian politics and bureaucracy combined with hints that national pride is at stake pose significant challenges.

Presently, our more visible actions are mainly legal prosecution, as allowed by the Transboundary Haze Pollution Act, passed in Parliament in 2014, and consumer action, which saw products taken off the shelves of supermarkets in 2015. Both of these are somewhat reactive and punitive in nature. Given the practical challenges to the former (as of March 2016, only two of six companies given notice by the National Environment Agency have responded) and limited impact of the latter due to the relatively small size of Singapore's market, the effectiveness of such measures is questionable.

An interesting development, probably unrelated but possibly opportune, is Singapore's National Research Foundation (NRF) committing 10 million Singapore dollars to Wilmar International, on a matching basis, to invest in technology startups based in Singapore. For Wilmar, the focus would be on "technologies that have the potential to cause a wide disruptive impact on agriculture, food, human and animal health, and industrial biotechnology." Certainly, Wilmar's business interests are diverse but it does have almost a quarter of a million hectares of planted oil palm, most of which is in Indonesia. Is it conceivable that part of this funding be used to invest in startup companies with innovative technologies or ideas in either tackling the root causes of the haze or more directly, in detecting and managing fires? Also, could such a model, where the government makes funds available — directly or through tax or other incentives — to parties which have the capabilities, ideas or potential to create solutions and/or devise innovative approaches to make a difference, be expanded?

Furthermore, while much is already known about the problem, could funding also be channelled to research to expand this body of knowledge, especially by objective and/or neutral parties or institutions like universities? Much could be gained, for example, in scientifically verifying whether the haze and fires have indeed become worse with time and if so what are the underlying causes. Besides environmental factors, it is likely that there are social, economic, political and legal drivers which have permitted or enabled a set of conditions which have led to the current state of affairs and/or the difficulty in addressing the issues. Understanding the complex interactions of these drivers via interdisciplinary research would inevitably help in the formulation of effective and sustainable solutions.

Beyond large companies and government, can ground-up community groups or people movements make an impact other than by driving consumer action? Taking the cue from what some of the larger companies are doing, there could be more opportunities for innovative actions in working with communities in tackling the root causes to prevent the occurrence of fires in the first place. There are already community-driven initiatives which are attempting to do this. A project by Majulah Community uses the crowdfunding platform Indiegogo to raise funds to help protect the Leuser ecosystem in Sumatra, recognising that it provides ecological services to Singapore. While framed primarily as a 'replanting' project, the organizers realise that communities, especially resident communities in the target areas, need to be involved. Besides working with them to restore damaged forests, they will also be focusing on trying to improve livelihoods in a sustainable manner. However, the ambitions of this particular group seem extremely modest given the initial funding target of 10,000 Singapore dollars. Still it is a start, which possibly also illustrates that besides having a general idea, most people do not know what is really required. Only by engaging will we find out and perhaps then can begin to do more.

Given the scale of the problem, the real and actual impact of such community-driven initiatives, and indeed even company or government-driven projects by themselves does seem questionable. It is not likely that the haze problem will be solved completely any time soon. Besides scale, whether or not and the extent to which any haze affect the region are also subject to natural forces not within our control.

However, it is important that government, companies, industry organizations like the RSPO, non-governmental organizations (NGOs) and communities — resident as well as those in neighboring countries — continue to seek ways to tackle the problem. It is very possible that the cumulative impact of the many current, planned and possible initiatives by these different stakeholders, using existing and developing technologies, innovative methods and increased knowledge, would reduce the extent and impact of transboundary haze. Indeed if these initiatives target the root causes of the problem, the benefits would go beyond the absence of haze. Forests, biodiversity and natural landscapes would be restored, preserved and protected. And communities' sustainable livelihood options would be developed and improved. Even by themselves, these would be excellent outcomes.

9 A New Thinking to Cooperation in Tackling the Indonesian Haze?

Kheng Lian Koh

Emeritus Professor, Faculty of Law,
National University of Singapore
Honorary Director,
Asia-Pacific Centre for Environmental Law (APCEL),
Singapore
lawkohkl@nns.edu.sg

Some 25 years have passed since the Indonesian haze (Indo haze) was first spotted. Subsequent episodes have caused widespread pollution not only in Indonesia but to other Association of Southeast Nations (ASEAN) member states, namely, Singapore, Malaysia, Brunei, Philippines and Thailand. The transboundary impact has resulted especially in economic losses and health risks. As for Indonesia, there have been serious disruption of the functioning of some of its communities, health issues and, more recently, loss of lives of its citizens, not to mention environmental losses.

Much ink has flowed on how to tackle the Indo transboundary haze. This includes ASEAN instruments such as the ASEAN Agreement on Transboundary Haze Pollution, 2002 (AATHP) and more recently, Singapore's Transboundary Haze Pollution Act, 2014 (STHPA). Literature on the haze abounds from the perspective of legal and other disciplines. Today nothing much has changed although Indonesia has adopted a multi-prong approach and brought some cases to courts. The recent announcement by the Indonesian Prime Minister Joko Widodo on 25 May 2016 to stop granting licences for palm oil plantations is a significant step but still awaits effective implementation.

In the meanwhile, Singapore's efforts to bring a director of an Indonesian company to a Singapore court under the STHPA on a suspected involvement in the 2015 forest fires has caused tensions between the two countries: "Indonesia reviewing collaborations with Singapore over haze" (*The Straits Times*, 16 May 2016: A6). Indonesia appears to be embarrassed on grounds of a "derogation of sovereignty", including its "national dignity and pride". Whether such response is justified or not underlies the need for what has been described as a "new paradigm needed from Indonesian ties with S'pore" (*The Straits Times*, 24 May 2016, A23); Jakarta 'will complete review of S'pore schemes next week' (*The Straits Times*, 25 May 2016) and "When diplomacy turns hazy" (*The Straits Times*, 25 May 2016). At the recent Second United Nations Environment Assembly (UNEA2), Senior Minister of State for Environment and Water Resources, Amy Khor raised concerns over the Indo haze pollution in achieving the UN 2030 Agenda for Sustainable Development, especially in promoting sustainable forest management: "Singapore raises concens over haze at UN meeting in Africa (*The Straits Times*, 29 May 2016). She called for greater regional and international cooperation.

Principle of sovereignty: Obstacle?

The crux of the problem raised by Indonesia is the principle of sovereignty and that of non- intervention.

Can there be a new way of thinking with a new approach to tackling the haze that would avoid the perceived derogation of sovereignty on the part of Indonesia? The answer is 'yes'. The author recommends framing the haze issue as 'disaster management' under the ASEAN Agreement on Disaster Management and Emergency Response, 2002 (AADMER, which complements AATHP); and also in ASEAN 2025: Forging Ahead Together (ASEAN 2025) under the ASEAN Political-Security Community (APSC). This new framing together with calls from the UN post-2015 SDGs (Sustainable Development Goals) and the UN 2030 Agenda, and other relevant instruments has, as its basis, an *obligation to cooperate (i.e., duty to cooperate)*. Such duty or *enhanced cooperation* will automatically calibrate the principle of sovereignty and that of non-intervention. An enhanced cooperation should not be viewed as a derogation of sovereignty. The above-mentioned instruments point to the importance of the duty to cooperate to achieve sustainable development. The haze with its 'destructive' force undermines sustainable development. The result of the new framing would strengthen the sum total of sovereignty of all the members states of ASEAN.

Indonesia has time and again unjustifiably applied the principle of sovereignty to its hilt in any attempt by Singapore and other countries to tackle the haze situation. This has posed an inverterate obstacle to cooperation. This principle is linked to the Treaty of Westphalia, 1648 when nations obtained independence from their colonial masters. As will be shown, it is an anachronistic principle in the context of environment in the broader sense and in the achievement of sustainable development.

As ASEAN moves forward as an integrated legal community, the words of the late S. Rajaratnam, then Minister of Foreign Affairs, Singapore, should be heeded. At the opening ceremony of the inaugural meeting of ASEAN Foreign Ministers in 1967 on the establishment of ASEAN, he cautioned:

> *It is easy to give birth to a new organization ... Now the really difficult task is to give flesh and blood to this concept .... The realization has grown and, therefore, it is necessary for us if we are really to be successful in giving life to ASEAN to marry national thinking with regional thinking. We must now think at two levels. We must think not only of our national interests but posit them against regional interests. That is a new way of thinking of our problems. And that is two different things and sometimes they can conflict. Secondly, we must also accept the fact, if we are really serious about it, that regional existence means painful adjustments to those practices and thinking in our respective countries. If we are not going to do that, then regionalism remains Utopia.*

This reasoning holds even more true today as the ASEAN Community forges ahead. Yet nationalism still runs high as Indonesia continues to perceive that its sovereignty is undermined if Singapore were to successfully bring an action against an Indonesian company director in a Singapore court under the STHPA. The ASEAN's integrated community should not be blown by a side wind of Indonesia's interpretation of sovereignty, its national pride, and a perceived affront to its dignity in this context.

Indonesia has the opportunity to take a leadership role in tackling the haze problem. It will strengthen "ASEAN 2025. Moving Ahead Together" and the UN post-2015 Sustainable Development Goals in achieving sustainable development.

Framing the issue as 'disaster management'

Framing the issue as one of 'disaster' could shift the paradigm.

The formal establishment of the ASEAN Community on 31 December 2015 brought about ASEAN 2025 which replaces the Roadmap for an ASEAN Community (2009–2015) in the Cha-Am Hua Hin Declaration on the Roadmap for an ASEAN Community (2009–2015). It continues to further strengthen the integration of ASEAN as a community. If ASEAN is to take this seriously, what does this entail for cooperation in tackling the Indo haze? Rather than harking on the Westphalian principle of sovereignty, which has the unintended consequence of steering away from cooperation, Indonesia should consider the haze issues from the viewpoint of 'environmental disaster' under the ASEAN Agreement on Disaster Management and Emergency Response, 2005 (AADMER). It entered into force on 24 December 2009, after ratification/acceptance by all ASEAN member states.

There is no doubt that AADMER applies to the Indo haze, having regard to the magnitude of the disruption. A 'disaster' is defined in Article 1 as "a serious disruption of the functioning of a community or a society *causing widespread human, material, economic or environmental losses.*" The objective of AADMER in Article 2 is to achieve substantial reduction of losses of lives and in social, economic and environmental losses. Under AADMER, there is what is termed the 'Receiving Party' defined as "... a Party that accepts assistance offered by an Assisting Entity or Entities in the event of a disaster emergency." A 'Requesting Party' is "... a Party that requests from another Party or Parties assistance in the event of a disaster emergency." An 'Assisting Entity' is "... a State, international organization, and any other entity or person that offers and/or renders assistance to a receiving party or a requesting party in the event of a disaster emergency."

There must be consent for assistance by the Receiving Party as well as the Requesting Party. Indonesia and all the other ASEAN member states have ratified AADMER and it came in force on 24 December 2009.

The AADMER was born out of the 2004 Indian Ocean tsunami in which Indonesia was greatly devastated. At that time, ASEAN was ill-prepared to deal with disasters but Indonesia was prepared to accept whatever assistance ASEAN member states including Singapore and other countries could render. It did not raise any issue of sovereignty.

It is not conceivable that if the haze is framed as a 'disaster', Indonesia would not seek assistance under AADMER. It did accept assistance during the Indian Ocean tsunami from Singapore and other countries even before AADMER. Would Indonesia decline any assistance offered under "disaster management" — it would sound churlish if it does.

Indeed, in the context of the impact of the haze on other ASEAN member states, there is a *duty to cooperate* under AADMER. Under Article 4(a), which uses *mandatory language*, when applied to the haze situation, Indonesia "... *shall* cooperate in developing the implementing measures to reduce disaster losses ..." Other relevant provisions of Article 4 include subsection (b) "... When the said disaster is likely to cause *possible impacts on other Member States*, the Parties *shall* ... respond promptly to a request for relevant information sought by a Member State or States that are or may be affected by such disasters, with a view to minimizing the consequences."

Furthermore, Article 4(c) *obliges* Parties to "promptly respond to a request for assistance by an *affected* Party." The affected Party/Parties in the Indo haze are not only Indonesia but its neighboring countries.

While there are provisions on sovereignty in AADMER (Article 3), the fact that consent is required for assistance does not raise any issue of its derogation. As will be seen (later), such assistance is framed as 'enhanced cooperation' under APSC in ASEAN 2025 in the securitization of disasters, and is not a traversty of the principles of sovereignty and non-intervention. In any case, the issue of sovereignty, if any, is not insurmountable.

In pursuing the objective of AADMER, Article 2 calls upon Parties to "*jointly respond* to disaster emergencies through concerted national efforts and *intensified regional and international cooperation.*"

It is puzzling that AADMER has been overlooked. Instead, the ASEAN member states (other than Indonesia) have patiently waited for Indonesia to ratify the AATHP while the haze has become a perenniel problem. Indonesia only ratified it in 2014 — 12 years after it came into existence.

Would not the Singapore Transboundary Haze Pollution Act, 2014 passed by Singapore (an affected ASEAN Party) not be viewed as complementary to the AATHP for effective enforcement? If so, where is the sting, if any? Is there a derogation of sovereignty, national pride or dignity?

The AADMER itself is based on the rationale that where the disaster cannot be managed by the affected state alone, nor by one region, the 'whole-of-the-world' is called upon to assist. This is demonstrated by the disaster when cyclone Yolanda made a landfall in the Philippines in 2014. Recognizing the 'whole-of-the-world' approach, ASEAN entered into a number of strategic plan of cooperation on disaster management with other organizations. An example is the Tripartite Memorandum of Cooperation (MoC) between ASEAN, the United Nations International Strategy and Disaster Risk Reduction (UNISDR) and the World Bank on Disaster Risk Reduction in the region in 2009. Indonesia itself recognizes this as it was part of ASEAN in the MoC. Also, the Cha-Am Hua Hin Statement on East Asia Summit (EAS) Disaster Management, 2009 reinforces the 'whole-of-the-world' approach.

One cannot imagine that faced with with a dire disaster, a state would refuse outside aid because it is seen to compromise its sovereignty, nationalism, pride, dignity and whatever, even if this is for the better good of the public.

AADMER goes further than AATHP which objective is only limited to the prevention and monitoring of the haze, and not a holistic approach in operationalization in reducing disaster and putting in place effective mechanisms.

The AADMER also incorporates the Sendai Framework on Disaster Risk Reduction, 2015 (SFDRR) adopted at the Third United Nations World Conference on Disaster Risk Reduction, held from 14 to 18 March 2015 in Sendai, Miyagi, Japan. The SFDRR replaces the Hyogo Framework for Action.

The emphasis of SFDRR is risk reduction in all the stages of disasters: pre, during and post. Two relevant principles of SFDRR are highlighted for ASEAN to better understand the importance of managing disaster risk under AADMER.

Priority 1. Understanding disaster risk.
Disaster risk management should be based on an understanding of disaster risk in all its dimensions of vulnerability, capacity, exposure of persons and assets, hazard characteristics and the environment. Such knowledge

can be used for risk assessment, prevention, mitigation, preparedness and response.

Priority 2. Strengthening disaster risk governance to manage disaster risk. Disaster risk governance at the national, regional and global levels is very important for prevention, mitigation, preparedness, response, recovery, and rehabilitation. It fosters collaboration and partnership.

ASEAN 2025

The Indo haze also calls for consideration of the ASEAN Political-Security Community under ASEAN 2025 (APSC 2025).

A 'disaster' which is classified under the ASEAN Socio-Cultural Community (ASCC, also contained in ASEAN 2025, i.e., the third pillar) can morphed into APSC which has a provision on 'comprehensive security' under the Non-Traditional Security (NTS) approach — this is the the first pillar. The two pillars are mutually reinforcing. The APSC 'securitizes' disaster management when it is sesious. The rationale of the NTS approach is human security. The NTS approach also reinforces *'enhanced* cooperation' such as having joint task forces, joint partnerships, entering the territory to deal with a disastrous situation. This is in line with ASEAN integration.

The relevant provisions of APSC under ASEAN 2025 are:

B.3.8. Strengthen ASEAN cooperation on disaster management and emergency response.

   (i) *Enhance joint effective and early response at the political and operational levels in activating the ASEAN disaster management arrangements to assist affected countries in the event of major disasters;*
  (ii) *Implement the ASEAN Agreement on Disaster Management and Emergency Response (AADMER) as the main common platform for disaster management in the region; and*
 (iii) *Ensure that disaster risk reduction is integrated into ASEAN strategies on disaster management and emergency response.*

B.2. Respond to urgent issues or crisis situations affecting ASEAN in an effective and timely manner.

  (i) Support the Chair of ASEAN in ensuring an effective and timely response to urgent issues or crisis situations affecting ASEAN, including providing its good offices and such other arrangements to immediately address these concerns;

  (ii) Convene special meetings at the Leaders, Ministers, Senior Officials or CPR levels, including through video conferencing in the event of crisis situations affecting ASEAN;

(iii) Activate the ASEAN Troika to address urgent situations affecting regional peace and stability in a timely manner;

(iv) Explore ways and means or applicable mechanisms which could be activated immediately to address urgent situations affecting ASEAN as well as regional peace and stability; and

  (v) Build on existing mechanisms to enhance early warning capability to prevent occurrence or escalation of conflicts.

The disastrous effects of the Indo haze which have been on for many years with periods of respite should, nonetheless, be 'securitized' under APSC. It should attract the non-traditional security (NTS approach which is based on human security).

Way forward

Finally, let us adopt the Earth Charter as an ethical foundation for a just, sustainable, and peaceful, global society. It starts with a powerful Preamble:

We stand at a critical moment in Earth's history, a time when humanity must choose its future. As the world becomes increasingly interdependent and fragile, the future at once holds great peril and great promise. To move forward we must recognize that in the midst of a magnificent diversity of cultures and life forms we are one human family and one Earth community with a common destiny. We must join together to bring forth a sustainable global society founded on respect for nature, universal human rights, economic justice, and a culture of peace. Towards this end, it is imperative that we, the peoples of Earth, declare our responsibility to one another, to the greater community of life, and to future generations.

As ASEAN forges ahead and is people-centered, rhetoric must be set aside for shared vision of human security, particulary at this critical period of man's history when what is needed is a focus on human security which brings about ethical values, equity, for a changing world with all manner of hazards — political, social, economic that undermine and threaten humanity. The UN 2015 SDGs and 2030 Agenda together with ASEAN 2025 should work in tandem to achieve sustainable development.

A reminder that under Article 2 of AADMER the objective is to pursue *"the overall context of sustainable development"*. This surely aligns AADMER with AATHP, UN 2015 SDGs and 2030 Agenda, ASEAN 2025 and the underlying ethics of the Earth Charter. Since the haze started over 25 years ago environmental law, policy and governance, not to mention international environmental law have seen tremendous changes. Indonesia should awaken to the fact that the Westphalian principle of sovereignty and that of non-intervention are anachronistic in the context of today's environment and must be calibrated according to context. It is not a departure of the ASEAN Way.

References

(2016, May 26). When diplomacy turns hazy. *The Straits Times*. Retrieved http://www.straitstimes.com/opinion/st-editorial/when-diplomacy-turns-hazy

Arshad, A. (2016, May 16). Indonesia reviewing collaborations with Singapore over haze. *The Straits Times*. Retrieved http://www.straitstimes.com/asia/se-asia/indonesia-reviewing-collaborations-with-singapore-over-haze

Arshad, A. (2016, May 25). Jakarta 'will complete review of S'pore schemes next week'. *The Straits Times*. Retrieved http://www.straitstimes.com/asia/se-asia/jakarta-will-complete-review-of-spore-schemes-next-week

Nugroho, J. (2016, May 24). New paradigm needed for Indonesian ties with Singapore. *The Straits Times*. Retrieved http://www.straitstimes.com/opinion/new-paradigm-needed-for-indonesian-ties-with-singapore

Tan, A. (2016, May 29). Singapore raises concerns over haze at UN meeting in Africa. *The Straita Times*. Retrieved http://www.straitstimes.com/singapore/environment/singapore-raises-concerns-over-haze-at-un-meeting-in-africa

10 The Palm Oil Industry and Southeast Asia's Haze: Outline for a Class Discussion

Linda Y. C. Lim

Professor of Strategy, Stephen M. Ross School of Business,
University of Michigan
lylim@umich.edu

Motivation

This class discussion examines the causes and impacts of, and proposed solutions to, Southeast Asia's transboundary 'haze' problem, as they relate to multiple stakeholder interests in the palm oil industry. It can fit into business, public policy or social science course modules focused on economic development, global supply chains, corporate social responsibility, sustainability, non-governmental organization (NGO) advocacy, and ethical investing. The goal is to show the multifaceted complexity of the problem, and hence the difficult challenges of devising effective solutions.

Suggested readings

Any relevant media articles can be used as bases for discussion which include:

"A green light for red palm oil as health aid?"
Wall Street Journal August 3, 2015
"Palm-oil migrant workers tell of abuses on Malaysian plantations"
Wall Street Journal July 26, 2015
"Southeast Asia, choking on haze, struggles for a solution"
New York Times October 8, 2015

"Norway oil fund excludes companies on environment risks"
   Wall Street Journal August 17, 2015
"Palm oil king goes from forest foe to buddy in deal with critics"
   Bloomberg Business Week March 12, 2015

The following case study can also be referred to:

Of Orangutans and Chainsaws: Cargill Inc. Confronts the Rainforest
   Action Network Advocacy (Ivey W12080).

Discussion

What causes the haze?

The haze is caused by burning of forest in Indonesia to replant the land with crops, of which palm oil is the highest value and the most common. When there is no rain to put out the fires, for example, because of the warm current El Niño in the Pacific Ocean, and given the peatlands under the forest, the fires can burn out of control, spewing wood ash and particles into the air. These are carried north and east above the equator by prevailing winds.

Who benefits from palm oil production and consumption?

The **end-consumer** is the chief beneficiary, because this is a cheap vegetable oil, relatively healthy, with no trans fats. (Palm oil enjoyed a big bonanza when the US Food and Drug Administration banned trans fats.) It preserves the shelf life of food and other products, like toiletries and cosmetics.

Corporate buyers like Unilever and Nestle in Europe, Procter & Gamble, Kraft, General Mills and Burger King in the US, and similar consumer product companies in Asia, also benefit.

Producers in Indonesia and Malaysia, who are both large plantations, some of them state-owned, and smallholders.

The **producing country** benefits from export earnings and taxes.

Workers employed in the industry benefit from the jobs and incomes earned.

The bottom line is that palm oil enjoys a very big market which is fast-growing, especially in Asia. As people in China and India get richer, they eat more processed foods, use toiletries etc.

Who loses from palm oil production, and why?

This is a classic case of an externality: harm is caused to those who do not produce the product. (Nearly everyone consumes it.)

Everyone loses from the global climate change to which palm oil contributes, but especially those who suffer directly from the resulting haze, mostly in Southeast Asia. Besides health problems, some will also suffer economic losses from reduced revenues and jobs from tourism and outdoor recreation.

Those who suffer the most from climate change are low-lying Pacific islands from rising sea levels, the Philippines from increased and more severe typhoons, Africa from drought etc. Since these are mostly low-income developing countries, there is a negative distributional consequence.

How does palm oil cultivation contribute to global climate change?

The cheapest and fastest way to clear land for palm oil (or other crop) cultivation is by burning the dense tropical forest, the ash from which also helps enrich the soil. 'Slash-and-burn' or shifting cultivation has been the traditional method of clearing the forest, including by indigenous peoples.

But in Indonesia, the forest is underlain by peat bogs (coal) which continue burning underground after a fire, unless heavy rain douses it. Global warming/climate change has made the rains more erratic, while expanded cultivation, given strong demand for palm oil, has vastly expanded the acreage to be cleared.

Removing the rainforest reduces carbon absorption by trees, which contributes to climate change and global warming. But this is the same for any use of former forest land, for example in Brazil for cattle farming, elsewhere for cities and industrial zones. Oil palm is less bad than these other uses because at least it replaces the forest with palm trees, and is renewable.

Removing the rainforest also causes loss of botanical and animal species, and loss of habitat for indigenous peoples. Why should we care?

Botanical and animal species might have future benefit for the pharmaceutical industry or scientific research.

There might be unanticipated consequences for the natural balance of life on earth, for example if particular species important for disease control (like natural predators of disease vectors) or plant/agricultural growth (like bees) are wiped out.

Utilitarian and commercial reasons are not the only ones. Environmentalists believe biodiversity is a good in itself that we should be willing to pay for through reduced consumption or taxation.

What about indigenous peoples? Most forest dwellers who live their traditional lifestyles have short, insecure and miserable lives. Is it not better for them to leave the forest and become modernized and educated so they have a chance at a better life? The counter-argument is it is not up to us to make a value judgment on other people's lives, which are also valuable for their own sake. But isn't it equally bad if they survive only because of curiosity value as tourist objects?

Isn't there a cost to forest preservation as well?

Yes, there is an opportunity cost in terms of lost jobs and income from timber harvesting and from farming and plantation labor, for which it has been argued that countries with forests should be compensated.

This is another example of an externality — the beneficiaries of carbon absorption by forests (the world as a whole) do not have to pay for the benefit, but expect the forested countries to bear the (opportunity) cost.

How would compensation (for not destroying forests) be funded? By a global carbon tax? Do you have other ideas?

But how beneficial is palm oil cultivation to workers?

We read that they are exploited migrant workers, even victims of 'human trafficking', and labor under horrible working conditions.

Not all palm oil workers are wage laborers on large corporate plantations. Many are smallholders who cultivate small plots, as owners or tenants.

Workers who 'choose' such employment usually do so because they do not have better alternatives in their home island or nation. They would be poorer if these jobs did not exist.

Why don't producers (large or small) employ more sustainable methods of cultivation (e.g., no burning) and offer better wages and working conditions?

Forests can be cut using capital-intensive technology (heavy equipment such as that used by loggers) but this is expensive and unaffordable for smaller producers.

Not burning would reduce air pollution from the haze, but not eliminate the other problems caused by deforestation, including loss of soil fertility from bare earth runoff.

So long as workers can be recruited under current conditions, there is no incentive or need to improve them.

Raising costs would reduce profits and the ability to attract capital from shareholders.

There is also concern that if competitors (whether of palm oil or substitute products) do not follow suit and bear the same costs, a producer could lose market share and revenues.

What is the responsibility of the Indonesian and Malaysian governments in the palm oil controversy and haze problem? What should they do?

Governments are responsible for regulations and their enforcement e.g., on land clearing, labor standards, pollution etc.

But governments may not have the incentive, political will or administrative capacity to enact and enforce the proper regulations. Business interests will lobby against regulations, and perhaps bribe to evade enforcement and punishment. Government monitoring and enforcement is difficult over Indonesia's very large, sprawling and fragmented cultivation area. The country is also highly decentralized, politically and administratively. The government itself may prefer to reap revenues from the industry rather than sustain costs policing it, given budget constraints.

If Indonesia does not regulate and enforce the industry, neither will Malaysia, since they compete on costs. Also, many plantations in Indonesia are owned by Malaysian companies.

Can the Malaysian government impose fines on Malaysian companies which are burning land in Indonesia? This is difficult without cooperation from the Indonesian government, since evidence of culpability must be obtained in Indonesia. Malaysian companies will also have political support within their own government to resist this. They would lose competitiveness against other companies like local Indonesian companies which do not have to bear the cost of such fines.

Plantations in Malaysia itself do not burn. This is because Malaysia is a smaller area, with well-established, mostly larger plantations, so they are better capitalized, more productive and easier to police. Malay smallholders are organized under the government agency Federal Land Development Authority (FELDA), so there is sharing of resources and standardization of practices. Farmers' children no longer want to be in the business, and labor is scarce (requiring the employment of mostly Indonesian migrant workers), so some farmers are selling out and plantations are consolidating.

Large plantations argue that fires are not set by them but by small farmers who move into newly logged forest cleared by lumber companies. These small farmers often cluster around the fringes of large plantations with which they share resources, like transport and processing facilities, so fires can spread easily to the large plantations, who give this explanation when satellite photos show their land burning.

Is there anything the Singapore government can do?

The Singapore government can provide technical and financial support to Indonesia for fighting fires once they have started. But this requires cooperation of the Indonesian government which might have 'sovereignty', security or national pride concerns in allowing a foreign government (most likely its military) to operate in its territory and airspace. Firefighting is also not very effective when fires have spread widely and conditions are dry, given the large area covered.

The Singapore government can impose large fines on Singapore-registered companies that satellite imagery shows have fire on their plantations in Indonesia. This faces the same problems noted above for the Malaysian government, plus could undermine Singapore's palm oil- and commodity-related trade and financial services sectors like trade finance, insurance and stock exchange listing as potentially liable businesses seek to avoid its jurisdiction. This must be balanced against losses from the haze in other sectors of the Singapore economy, such as tourism and retail.

Is there anything end-consumers can do?

Consumer boycotts in general are difficult because (a) there are so many millions (in this case, billions) of unorganized consumers of any particular product, and (b) most of them are not informed of the situation and/or do not care about it enough to inconvenience themselves by changing and paying more for purchases.

Boycotting palm oil is particularly difficult because (a) it is in so many daily-use products (cooking oil, processed foods, toiletries, cosmetics), (b) it is in very small quantities in, and thus a small proportion of the value of, each product, (c) it is an intermediate, not a final product, so its presence is difficult to detect in the complex composition of a product, (d) it is a commodity, so it is difficult to differentiate particular producers, (e) there may be no substitutes without (f) expending a lot of effort and paying a lot more, and (g) it has many good qualities that substitutes might not possess e.g., as a food preservative and relatively healthy fat.

Supermarkets and consumers in Singapore have boycotted paper products produced by one very large palm-oil producer, Asia Pulp and Paper (APP), believed to have contributed to the haze. This was possible because Singaporeans were directly adversely affected by the haze, the company produced other (non-palm oil) products readily identified, consumers were aware of the connection, and many substitutes were available (that were not more expensive; if they were more expensive, Singapore's high-income consumers could easily afford them). Palm oil produced on APP estates could not be easily identified for boycott.

In contrast, the fastest-growing consumer markets for palm oil are in large developing countries like China and India, where lower-income, less-aware, less-affected and more price-sensitive consumers are less likely to practice boycotts.

What are NGOs (non-governmental organizations) doing in their efforts to encourage sustainability in the palm oil industry?

They are running campaigns to raise buyer and consumer awareness, hoping that the risk of loss of reputation/'brand hit' and market share will encourage big Western consumer good buyers to put pressure on their Southeast Asian suppliers to produce and sell only 'sustainably produced' palm oil.

They are strategically focusing on a few of the largest sellers and buyers which have the scale and market power to "change how the industry does business."

Cargill was targeted by the Rainforest Action Network (RAN) because it both buys and trades palm oil and operates plantations in Indonesia, thus covering the whole palm oil supply-chain. It is also a private company so could be less short-term focused than US publicly listed companies, and with fewer stakeholders might be able to take decisions and execute them faster.

A consortium of NGOs has targeted Wilmar and Unilever, respectively the world's largest supplier and buyer. Having them work together can help ensure that other industry players follow similar sustainable practices.

NGOs have joined other stakeholders (including investors, producers and consumers) in the Roundtable for Sustainable Palm Oil (RSPO), an international association that hopes that collectively setting standards will ensure compliance by all parties, since none can get an unfair competitive advantage by avoiding higher costs if those are necessary.

What are the limitations of this collective, negotiated approach?

Not all parties are members, for example myriads of smaller farmers in Indonesia will not have the means, interest or incentive to participate, yet they may be the ones with the most unsustainable practices, like clearing land by forest burning.

Compliance is voluntary, monitoring by professional auditors is expensive, and there are no penalties for non-compliance, so effectiveness might be limited.

Requiring buyers to vet suppliers for sustainability is likely to reduce the number of suppliers in favor of larger producers, penalizing smaller farmers. Sustainable may not be equitable.

Focusing on sustainability alone does not deal with the larger issue of deforestation, which should limit or reduce palm oil cultivation acreage.

The main actors pushing for sustainability are Western companies and NGOs, when the fastest-growing markets and increasingly large buyers are in lower-income developing countries, particularly China and India.

What else can be done, and by whom?

Buyers can take unilateral action to cut off suppliers guilty of unsustainable practices e.g., the Sinar Mas/APP group was dropped by Burger King, Carrefour, Unilever and Nestle. But there are always other buyers.

Norway's sovereign wealth fund has divested from most palm companies, and this seems to be what prompted Wilmar's boss to "get serious" about the situation. But more investors will have to do this to impose a cost to shareholders of unsustainable practices. The industry's growth and profitability make it likely that other investors will be interested.

Do you think the haze will continue to be a problem for the foreseeable future? Why or why not?

Open Discussion.

Extending the discussion

Students can be divided into groups to research and present detailed different solutions to the problem e.g., from the viewpoints of public policy, international relations (e.g., ASEAN efforts), corporate strategy and community activism.

11 Indonesia's Fires in the 21st Century: Causes, Culprits, Impacts, Perceptions, and Solutions

Erik Meijaard

Director, Borneo Futures
Honorary Director, Center of Excellence for Environmental Decisions,
University of Queensland
emeijaard@gmail.com

Introduction

Indonesia's 2015 fires

Indonesia's fires were coined the greatest environmental disaster of the 21st century (Meijaard, 2015; Koplitz *et al.*, 2016). Compared to other large environmental disasters, such as the British Petroleum oil spill in the Gulf of Mexico in 2012, the economic, social and environmental impacts associated with Indonesia's 2015 fires appeared to significant exceed impacts (Meijaard, 2016). Still, despite the magnitude and impact of the fire event, relatively little attention was paid to them in the global media. Why was this the case? An estimated 40–80 million people living directly downwind from Indonesia's fires were breathing in noxious fumes day in day out for months, hundreds of thousands of people fell ill, and many fire- and haze-related fatalities were reported. Furthermore, the fires were reported to be a significant economic cost to the Indonesian economy, estimated at 1.8 percent of the country's gross domestic product. Singaporean, Malaysian, and Indonesian people directly affected by the events complained and raised their concerns in local media, but considering the major regional and global impacts, it is surprising how little attention the disaster received outside the region.

In this chapter, I assess perceptions about Indonesia's fires, provide an overview of their impacts, discuss underlying drivers of the fires, and review potential policy interventions that might prevent future occurrences.

History of Indonesian fires

Forest and land fires are not only a recent phenomenon in Indonesia. Explorers to Borneo dating back to the 15th century reported forest fires, and such reports continued to the 20th century (Dennis, 1999). Evidence of ancient fires based on radiocarbon dating of charcoal in soils has been reported in various parts of the region (Goldammer and Seibert, 1989; Anshari *et al.*, 2000), indicating that fires, probably anthropogenic ones, did occur. The log books of Portuguese and Dutch explorers in Southeast Asia in the 15th and 16th centuries reported huge fires in peat swamps in southern Borneo, and a choking haze that extended as far as present-day Singapore.

Historical analysis and cross-referencing show that Southeast Asian fires occurred primarily in particular years that we now know are El Niño–Southern Oscillation (ENSO) years (Brookfield, 1997; Potter, 1997; Knapen, 2001). Throughout the 19th century, Singaporean newspapers reported smoke and haze problems in every ENSO year (Byron and Shepherd, 1998). In 1877, Bock (1881) reported a very serious drought along the coasts of Borneo and Sulawesi. The forests were reportedly dry and leafless, and there was a shortage of food and drinking water. Drought conditions lasted from March 1877 until March 1888, with only a little rain falling during the November and December 'wet season' (Knapen, 1997; Knapen, 2001). Michielsen (1882), who visited what is now Central Kalimantan Province in 1880, remarked on many locations seriously affected by large forest fires. Probably in reference to the fires of 1877: "It is hard to identify how the fire has spread through the forest. Although we had continued our way after leaving the peat swamp forest ... , we alternately passed dense forest, apparently unaffected by the fires, and completely burned patches, which even now [three years later], only showed sparse regrowth of shrub."

Other large fires in what is now western Indonesia were reported in 1846, 1853, and 1868 (Boomgaard *et al.*, 1997, pp. 150–151), adding further evidence that fires in this region are not a recent phenomenon only. It is also obvious that these fires had a significant impact on land cover. For example, various historic observations from the Danau Sentarum area in West Kalimantan (Indonesian Borneo) near the border with Sarawak indicate that fires burned extensive areas of forest (Gerlach, 1881; Molengraaff, 1900; Pfeiffer, 1993), and were probably an important factor in pre-industrial deforestation. In fact, many people seem to be unaware that deforestation for slash-and-burn agriculture had opened up significant parts of Borneo prior to the 1950s (Planning Department of the Forest Service, 1950; Knapen, 2001).

Despite Indonesia's deep fire history, the environmental, social and economic impacts of fires appear to have significantly increased in the late 20th century. The 1982/1983 fires were a major wake up call. From July 1982 to April 1983, an exceptional drought struck East Kalimantan and other parts of Indonesia. This coincided with a particularly marked ENSO event (Dennis, 1999). The fires started in November 1982 and increased in intensity and impact between January and April 1983. Estimates of the area burned vary, but probably up to 4.7 million hectares of land in Kalimantan alone burned, large parts of which were forest (Malingreau *et al.*, 1985; Leighton and Wirawan, 1986; Schindele *et al.*, 1989).

Subsequent to the major fires in 1982/1983, major forest and land fires became an increasingly frequent feature in Indonesia, recurring whenever major drought conditions prevailed. This included 1987, 1991, 1994 (when nearly 5 million hectares of land burned), 1997/1998 (when again several million hectares of land burned) (Dennis, 1999), 2002, 2006 (Langner and Siegert, 2009), 2009 (Yulianti *et al.*, 2012), and the earlier mentioned 2015 fires. It is safe to say that over the past few decades, fires have been a frequent feature of dry season landscapes in Indonesia, and few parts of the country remain unaffected (Figure 1).

Are forest fires a typical Indonesian problem?

Indonesia is obviously not the only country in the world experiencing fires. In fact, large parts of the vegetated tropics, sub-tropics, warm temperate, and boreal vegetation zones experience fires (Figure 2). What

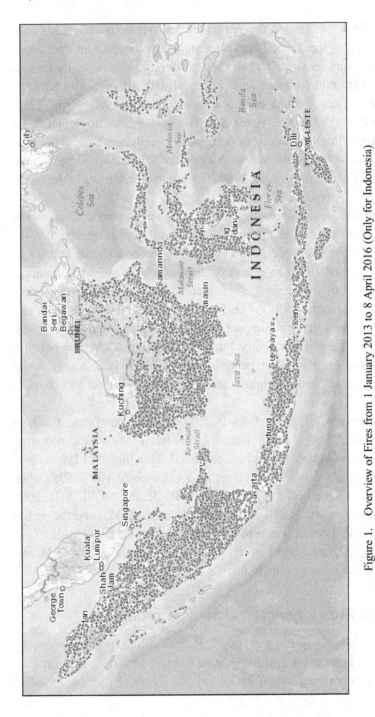

Figure 1.   Overview of Fires from 1 January 2013 to 8 April 2016 (Only for Indonesia)

Source: Global Fire Watch. The depicted size of each fire hotspot at the scale of all of Indonesia significantly exceeds the size of the actual fires, therefore the impression that nearly all land in Indonesia is burned is an artefact of the mapping scale and does not accurately reflect reality. What the map does show is how wide-ranging fires are in Indonesia, occurring in nearly all parts of the country.

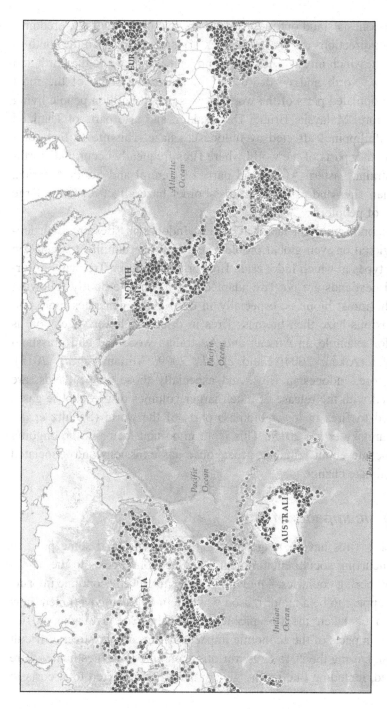

Figure 2. MODIS-Detected Active Fires between 8 and 14 April 2016

Source: Global Fire Watch.

makes Indonesia stand out in the world are two factors. Firstly, predominant wind directions during Indonesia's dry season are from the south to south-west, which means that smoke and haze are blown to the north and north-west. North and north-west of Indonesia lie some of the most densely populated parts of the world, with some 80 million people living in Singapore, Malaysia, Brunei Darussalam, and also parts of Thailand and the Philippines affected by Indonesia's haze. This means that compared to other parts of the world where fires frequently occur, like northern Australia, eastern Africa and parts of Central and South America, Indonesian fires and associated cross-border haze affect relatively large numbers of people.

A second important factor as to why Indonesian fires are a particularly regional or even global problem relates to the dominant vegetation and soil types in which Indonesian fires primarily occur. Tropical rainforests and peatlands are not fire adapted and regenerate poorly following fires. The moist biomass, especially in deep peats, generates more and more noxious haze then biomass fires in drier and lower vegetation as exists, for example, in African and Australian woodland and grassland savannahs (Aiken, 2004; Field *et al.*, 2009; Miriam *et al.*, 2015). Furthermore, Indonesian fires, and especially those on peatlands, are associated with the release of much larger volumes of greenhouse gases compared to fires in lower biomass parts of the world (Schultz *et al.*, 2008; Turetsky *et al.*, 2015). This is an important factor in the ongoing global debate about reducing greenhouse gas emissions and associated global climate change.

Economic impacts of fires

Indonesia's fires have a range of negative impacts (and some positive ones), incurring social, environmental and economic costs to the nation and neighboring countries. Among these, the perceived economic impacts of fires appear to be a major reason why the Indonesian government now seems to be taken the fire problem seriously (Soeriaatmadja, 2016). Accurate impacts of the economic impacts of fires are, however, difficult to obtain. During the 2015 fires, various economic impact estimates were published, including one by the government that indicated total costs to

Indonesia of 35 billion US dollars. Another reported government estimate at 16 billion US dollars is half that costs, but still suggests losses of up to 1.8 percent of Indonesia's GDP (The World Bank, 2015). To what extent these estimates are accurate is hard to judge and will require further study. In comparison it is useful to look at previous estimates of economic impacts, for example, those following the major fires in 1997 and 1998.

Glover and Jessup (1999) initially estimated the total costs of the 1997/1998 fires at over 4 billion US dollars (Table 1). Their analysis shows the wide-ranging impacts that fires and haze can have. Tacconi (2003) later revised these estimates and found that the 1997/1998 fires resulted in forest degradation and deforestation causing economic costs in the range of 1.62–2.7 billion US dollars, while the costs of smoke haze pollution were in the range of 674–799 million US dollars. The valuation of costs associated with carbon emissions indicates that these may amount to as much as 2.8 billion US dollars. Tacconi acknowledged that his estimates did not include costs to Indonesian businesses as reliable estimates for these impacts were not available.

The uncertainties in these estimates are indicated by another study of the economic costs of the 1997/1998 fires in Indonesia. Varma (2003) looked at welfare losses associated with losses of marketed goods, and losses of environmental goods and services and estimated total losses of 20.1 billion US dollars as a result of use of slash and burn in Indonesia in 1997–1998. This estimate also included the benefits of slash-and-burn practices — which are cheaper than manual land clearing — indicating that use of these techniques is highly inefficient from a socio-economic perspective. This study, however, uses few empirical data from the actual study region and it is unclear how well the estimates approach the real costs.

Despite the variation in estimates of the socio-economic impacts of major fire and haze events in Indonesian and neighboring countries, it is obvious that costs are significant, often multiple billion dollars, or up to two or more percent of Indonesia's 2015 GDP of nearly 890 billion US dollars. Also, the above estimates do not incorporate the losses of forest ecosystem services besides sequestered carbon. For example, only a few studies have looked at how wildlife is impacted by fires in Indonesia (Doi, 1986; Suzuki, 1991; Azuma and Wirawan, 1994; Cleary and Genner,

Table 1.   Summary of Indonesia's Haze and Fire Impacts, 1997

Impacts	Economic Costs (US$ Million)
1. *Haze impacts*	
Medical costs	294.70
Productivity	167.30
Indirect impacts	462.00
Sub-total of health impacts	**924.00**
Tourism impacts (maximum)	70.35
Airline impacts	7.54
Airport closures	10.00
Sub-total of tourism/airline/airport impacts	**87.89**
Total haze impacts	**1,011.89**
2. *Fire impacts*	
Timber losses	493.67
Agriculture/plantation losses	470.39
Direct forest ecosystem production losses	704.97
Indirect forest ecosystem function losses	1,077.09
Domestic (capturable) biodiversity losses	30.00
Fire-fighting costs	11.67
Sub-total of impacts on Indonesia	**2,787.79**
Carbon release	272.10
Global biodiversity losses	not estimated
Fire-fighting costs	13.46
Sub-total of global impacts	**285.56**
Total fire impacts	**3,073.36**
Total overall impacts	**4,085.25**

Source: Glover and Jessup (1999).

2004; Slik and Van Balen, 2006), and it remains impossible to extrapolate findings from such localized studies to the larger areas affected by fires and haze. Such impacts could include reduced insect densities because of drought, haze and fire, and changes in the timing of flowering and fruit setting, resulting in reduced crop yields and production of products like honey. Similarly, it is unclear how populations of species like Bearded Pigs *Sus barbatus* which undergo population expansions in reaction to strong mast fruiting events are affected by potential changes in fruit availability induced by fire and haze. These pigs are the biggest source of protein to communities in the interior of islands like Borneo, and the value of free bushmeat is worth several billion US dollars per year. Many such potential impact pathways remain unstudied but if negatively affected these could have significant economic impacts beyond those estimated in the studies above.

Who is responsible for Indonesia's fires?

One major area of contention regarding Indonesia's fires is culpability. In fact, the blame game surrounding Indonesia's annual fire and haze problem is probably as old as the fires themselves (Carrasco, 2013). In 2015, for example, the public, politicians and media were quick to point the finger at large plantation companies, who have in the past been associated with land clearing using fire (Dennis *et al.*, 2005; Dennis and Colfer, 2006). More recent studies of fire and haze in Kalimantan and Sumatra, however, point towards small-scale farmers and other under-the-radar, mid-scale landowners, rather than large companies as the main cause of fires and haze. One study in Sumatra showed that 59 percent of fire emissions originated from outside timber and oil-palm concession boundaries. These non-concession-related fires generated 62 percent of smoke exposure in equatorial Southeast Asia (primarily Singapore and Malaysia) (Marlier *et al.*, 2015). In Kalimantan, non-concession fires play an even bigger role, with fires outside concessions generating 73 percent of all emissions and 76 percent of smoke affecting equatorial Southeast Asia (Marlier *et al.* 2015). These findings are in line with similar results based on more detailed studies in Riau Province (Sumatra). In Riau, 52 percent of the total burned area in 2013 was within concessions. However, 60 percent of these burned areas were occupied and used by small and medium landholders (Gaveau *et al.*, 2014). Another scientific publication on the causes of Indonesian forest fires showed that, even 15 years ago, when oil palm companies were involved a lot more frequently than now in land clearing with fire, small and medium scale land clearing, including for land speculation or resulting from land use conflicts, were a major cause of fires in both Sumatra and Kalimantan (Dennis *et al.*, 2005).

It appears from these findings that the simplified picture of big industries being the primary driver behind the fire and haze problems with communities playing a minor role at best through traditional burning methods, is not in line with reality and requires more nuanced analysis and discussion. The use of fire for land clearing is prohibited by Indonesian law in the industrial plantation context, but burning of a maximum of 2 hectares per household was legally allowed up to early 2016 (new laws are considered to reduce this). Data from the Indonesian Bureau of Statistics (BPS) *Potensi Desa* (village potential) surveys in 2014 showed that out of 7,191 villages in Kalimantan, 4,127 (57 percent) reported that communities in these villages had used fire in the previous year for slash-and-burn cultivation (*ladang* and *kebun*). The BPS data from 2008, suggest that those villages that did burn in 2007, burned on average 96.2 hectares (SD = 524) per village in the three provinces for which data were available (East, Central, and South Kalimantan). This indicates that some 397,000 hectares of secondary forest and scrub land were burned for small-holder agriculture in Kalimantan in 2007. In Sumatra, the figure is lower. The BPS data from 2008 covering seven provinces indicate that 3,152 out of 21,487 (= 14.7 percent) burned land for small-holder agriculture in 2007, with an average burned area of 52.8 hectares (SD = 839) per village. This would imply a total area of 166,394 hectares burned in Sumatra in 2007 for small-holder agriculture. This information only refers to one year, 2007, and it is unclear how representative it is for other fire years. Moderate La Niña conditions prevailed in 2007 (http://ggweather.com/enso/oni.htm), making it a wetter year than normal, and it is safe to assume that the cumulative reported burned areas for Kalimantan and Sumatra in 2007 (563,394 hectares) are less than in years with high fire activity when the conditions for opening up land would be more favorable. Half a million hectares of land annually burned for slash-and-burn agriculture in Sumatra and Kalimantan might thus be a conservative estimate. Considering that over a 47-year time frame (1960–2006), about 1 million hectares of land burned in Indonesia per year, with 90 percent of these burned areas in Kalimantan and Sumatra (Heil, 2007), it is likely that at least half the areas burned annually are burned by small-holders. This is also in line with the above studies that show that more than half of recent fires in Kalimantan and Sumatra could not directly be attributed to industrial-scale plantations. Thus the cumulative impacts of burning by small and medium land holders are a major contributor to the fire and smoke problem.

Irrespective of whether fires were initially lit by companies or smallholder farmers, there is another important factor that requires specific attention from government authorities if fire frequency and severity are to be reduced. Land use and tenure are subject to a bewildering range of laws and regulations that apply at different levels of government (national, province, district). It is often not entirely clear to whom a particular piece of land belongs, and whether underlying land claims are formal (based on permits) or informal (based on claims or long-term use). In that context, fire is a useful tool to make informal claims on a particular piece of land where tenure remains unresolved. Traditionally, lands opened up with fire and planted establish a certain level of informal ownership. Such claims can be used to start securing land for future more permanent use, or, if another entity or person wants the land, to increase the value of land and thus the price for which it is sold. Such speculative use of fire for claiming land rights remains poorly studied, also because the process is often not fully legal. A better understanding of why people burn in certain land use contexts is needed to reduce the incidence of fires, and also to provide some clarity in the ongoing blame game inherent to the Indonesian fire problem (Dennis *et al.*, 2005).

Discussions

There is growing recognition that forest fires are a global concern, both from a point of commitments to reduce greenhouse gas emissions and also to reduce air pollution and loss of forests. The global burned area has slowly been decreasing between 2000 and 2012 at a rate of 4.3 million hectares per year (-1.2 percent), but annual burned areas were still increasing in Southeast Asia over that time frame (Giglio *et al.*, 2013). With Indonesia's major fires in 2015, this latter trend likely further increased. It is therefore no surprise that Indonesia's 2015 fires were an important point of discussion in the United Nations 21st Conference of Parties in Paris, France in December 2015. Reducing Indonesia's major greenhouse gas emissions requires better forest management and prevention of future fires, especially in peatlands that store large quantities of carbon (Hooijer *et al.*, 2006). Indonesia seems to heed the global and regional attention on its fire problem and following the 2015 fires and haze disaster in Southeast Asia, there appears to have been a significant shift in political views in Indonesia as to how the fire issue should be managed. From an attitude of considering fires a necessary part of the development, to be controlled when excessive but otherwise not a major issue, the Government of Indonesia is now indicating that significant regulatory and budgetary steps will be taken to prevent an event similar to 2015 from re-occurring. None of these measures have yet been confirmed or signed off into new law, but they reportedly include a total fire ban, including for previously sanctioned community burning, and holding local government officials accountable for fires, punishing those that did not prevent them and rewarding those that did.

Banning fires altogether, including traditional burning practices, will require much more than new laws and their enforcement. Many of the most marginalized people in countries like Indonesia remain dependent on cheap methods of land clearing and fertilization using fire. Replacing clearing by fire-stick with manual or machine methods will require a significant change in agricultural practices and land use mind-sets, and likely major financial and technical support from the government to help farmers transition to new, more productive forms of land use. The challenge is particularly large in peat lands that have very low soil fertility, large numbers of poor farmers, and are highly prone to fires. Alternative forms of agriculture that provide people with a decent living will be hard to develop and if better peat protection is the objective, permanent tree crops would need to be developed that can grow in undrained conditions. Indonesia's recent announcement of a total ban on the issuing of plantation licenses on peatlands indicates that the government is thinking in that direction, but effective implementation will be challenging.

Acknowledgments

I thank the Arcus Foundation for supporting the Borneo Futures initiative.

References

Aiken, S. R. (2004). Runaway fires, smoke-haze pollution, and unnatural disasters in Indonesia. *Geographical Review*, 94: pp. 55–79.

Anshari, G., Kershaw, A. P. and van der Kaars, S. (2000). A late Pleistocene and Holocene pollen and charcoal record from peat swamp forest, Lake Sentarum Wildlife Reserve, West Kalimantan, Indonesia, *Palaeogeography, Palaeoclimatology, Palaeoecology*, 171: pp. 213–228.

Azuma, S. and Wirawan, N., editors. (1994). *Early Recovery Process of Kutai Ecosystem: A Preliminary Report*, Kyoto: Kyoto University Primate Research Centre.

Bock, C. (1881). *The Headhunters of Borneo: A Narrative of Travel up the Mahakam and down the Barito, Also Journeyings in Sumatra*, London, UK: Sampson Low.

Boomgaard, P., Colombijn, F. and Henley, D. (1997). *Paper Landscapes. Exploration in the Environmental History of Indonesia*. Leiden, The Netherlands: KITLV Press.

Brookfield, H. (1997). Landscape history: Land degradation in the Indonesian region, in Boomgaard, P., Colombijn, F., Henley, D. (eds.), *Paper Landscapes Explorations in the Environmental History of Indonesia*, Leiden, The Netherlands: KITLV Press, pp. 27–60.

Byron, N. and Shepherd, G. (1998). *Indonesia's Fire Problems Require Long Term Solutions*. London, UK: Overseas Development Institute.

Carrasco, L. R. (2013). Silver lining of Singapore's haze, *Science*, 341: pp. 342–343.

Cleary, D. F. R. and Genner, M. J. (2004). Changes in rain forest butterfly diversity following major ENSO-induced fires in Borneo. *Global Ecology & Biogeography*, 13: pp. 129–140.

Dennis, R. A. (1999). A review of fire projects in Indonesia 1982–1998. Center for International Forestry Research: Bogor, Indonesia, p. 112.

Dennis, R. A. and Colfer, C. P. (2006). Impacts of land use and fire on the loss and degradation of lowland forest in 1983–2000 in East Kutai District, East Kalimantan, Indonesia, *Singapore Journal of Tropical Geography*, 27: pp. 30–48.

Dennis, R. A., Mayer, J., Applegate, G. *et al.* (2005). Fire, people and pixels: Linking social science and remote sensing to understand underlying causes and impacts of fires in Indonesia, *Hum Ecol*, 33: pp. 465–504.

Doi, T. (1986). Present status of the large mammals in the Kutai National Park, after a large scale fire in East-Kalimantan, Indonesia. Kagoshima, Japan: Kagoshima University.

Field, R. D., van der Werf, G. R. and Shen, S. S. P. (2009). Human amplification of drought-induced biomass burning in Indonesia since 1960, *Nature Geoscience*, 2, 185–188.

Gaveau, D. L. A., Salim, M. A., Hergoualc'h, K. *et al.* (2014). Major atmospheric emissions from peat fires in Southeast Asia during non-drought years: Evidence from the 2013 Sumatran fires, *Scientific Reports*, 4, 6112.

Gerlach, L. W. C. (1881). Reis naar het meergebied van den Kapoeas in Borneo's Westerafdeeling. Met naschrift door Robidé van der Aa, *Bijdragen tot de Taal-, Land- en Volkenkunde van Nederlandsch-Indië Vierde volgreeks*, 5.

Giglio, L., Randerson, J. T. and van der Werf, G. R. (2013). Analysis of daily, monthly, and annual burned area using the fourth-generation global fire emissions database (GFED4), *Journal of Geophysical Research: Biogeosciences*, 118: pp. 317–328.

Glover, D. and Jessup, T. (1999). *Indonesia's Fires and Haze: The Costs of Catastrophe*, Singapore and Ottawa, Canada: Institute of Southeast Asian Studies and International Development Research Centre.

Goldammer, J. G. and Seibert, B. (1989). Natural rain forest fires in Eastern Borneo during the Pleistocene and Holocene, *Naturwissenschaften*, 76: pp. 518–520.

Heil, A. (2007). Indonesian forest and peat fires: Emissions, air quality, and human health, Ph.D. thesis. p. 142. Hamburg, Germany: Max Planck Institute for Meteorology.

Hooijer, A., Silvius, M., Wösten, H. and Page, S. (2006). PEAT-CO_2. Assessment of CO_2 emissions from drained peatlands in SE Asia, Delft Hydraulics report Q3943, Delft, the Netherlands: Delft Hydraulics.

Knapen, H. (1997). Epidemics, droughts, and other uncertainties on Southeast Borneo during the eighteenth and nineteenth centuries, in Boomgaard, P., Colombijn, F., Henley, D. (eds.), *Paper Landscapes Explorations in the Environmental History of Indonesia*, Leiden: KITLV Press, pp. 121–152.

Knapen, H. (2001). *Forests of Fortune? The Environmental History of Southeast Borneo, 1600–1880*, Leiden: KITLV Press.

Koplitz, S., Mickley, L., Marlier, M., Buonocore, J., Kim, P., Liu, T., Sulprizio, M., DeFries, R., Jacob, D. and Schwartz, J. (2016). Public health impacts of the severe haze in Equatorial Asia in September–October 2015: Demonstration of a new framework for informing fire management strategies to reduce downwind smoke exposure, *Environmental Research Letters*, 11(9).

Langner, A. and Siegert, F. (2009). Spatiotemporal fire occurrence in Borneo over a period of 10 years, *Global Change Biology*, 15: pp. 48–62.

Leighton, M. and Wirawan, N. (1986). Catastrophic drought and fire in Borneo tropical rain forest associated with the 1982–1983 El Nino southern oscillation event, in Prance, G. T. (ed.), *Tropical Rain Forest and World Atmosphere*, Boulder, Colorado: Westview Press, pp. 75-101.

Malingreau, J. P., Stephens, G. and Fellows, L. (1985). Remote sensing of forest fires: Kalimantan and North Borneo in 1982–83. *AMBIO*, 14: pp. 314–321.

Marlier, M. E., DeFries, R. S., Kim, P. S. *et al.* (2015). Fire emissions and regional air quality impacts from fires in oil palm, timber, and logging concessions in Indonesia, *Environmental Research Letters*, 10, 085005.

Meijaard, E. (2015, October 24). Indonesia's fire crisis — The biggest environmental crime of the 21st century. *The Jakarta Globe*. Retrieved http://jakartaglobe. beritasatu.com/opinion/erik-meijaard-indonesias-fire-crisis-biggest-environmental-crime-21st-century/.

Meijaard, E. (2016). Slash and burn practice in agriculture development in Indonesia. ICOPE Conference: Sustainable Palm Oil and Climate Change: The Way Forward Through Mitigation and Adaptation, Affiliation: Nusa Dua, Bali, Indonesia, p. DOI: 10.13140/RG.13142.13141.13243.19448.

Michielsen, W. J. M. (1882). Verslag eener reis door de boven districten der Sampit en Katingan rivieren in Maart en April 1880, *Ind Taal-Land-en Vdhenkuncle*, Tyds v: pp. 1–87.

Miriam, E. M., DeFries, R. S. and Kim, P. S. (2015). Fire emissions and regional air quality impacts from fires in oil palm, timber, and logging concessions in Indonesia, *Environmental Research Letters*, 10, 085005.

Molengraaff, G. A. F. (1900). *Borneo-Expeditie. Geologische verkenningstochten in Centraal-Borneo*, Leiden and Amsterdam: Boekhandel en Drukkerij E.J. Brill.

Pfeiffer, I. (1993). *Abenteuer Inselwelt: Die Reise 1851 durch Borneo, Sumatra und Java (editted by Gabriele Habinger)*. Originally published as *Meine Zweite Weltreise, Wien, 1856*, Himberg, Austria: Wiener Verlag.

Planning Department of the Forest Service (1950). Vegetation map of Indonesia. Scale 1: 2500000. Under supervision of L. W. Hannibal.

Potter, L. (1997). Where there is smoke there's FIRE, *Search*, 26: pp. 307–311.

Schindele, W., Thoma, W. and Panzer, K. (1989). Investigations of the steps needed to rehabilitate the areas of East Kalimantan seriously affected by fire: The forest fire 1982–83 in East Kalimantan, part I: The fire, the effects, the damage, the technical solutions, FR-report No. 5. Deutsche Forstservice GmbH: Jakarta, Indonesia.

Schultz, M. G., Heil, A., Hoelzemann, J. J. *et al.* (2008). Global wildland fire emissions from 1960 to 2000, *Global Biogeochemical Cycles*, 22: n/a-n/a.

Slik, J. W. F. and Van Balen, S. (2006). Bird community changes in response to single and repeated fires in a lowland tropical rainforest of eastern Borneo, *Biodiversity & Conservation*, 15: pp. 4425–4451.

Soeriaatmadja, W. (2016, January 19). Forest fires: Jokowi warns local officials. *The Straits Times*. Retrieved http://www.straitstimes.com/asia/se-asia/forest-fires-jokowi-warns-local-officials.

Suzuki, A. (1991). Forest fire effects on the population of primates in Kutai National Park. East Kalimantan, Indonesia, in Ehara A. (ed.), *Primatology Today*, Elsevier Science Publishers B. V. (Biomedical Division), pp. 51–54.

Tacconi, L. (2003). *Fires in Indonesia: Causes, Costs and Policy Implications*, Bogor, Indonesia: CIFOR.

The World Bank (2015). Indonesia's fire and haze crisis. Retrieved http://www.worldbank.org/en/news/feature/2015/12/01/indonesias-fire-and-haze-crisis. Accessed on 6 May 2016.

Turetsky, M. R., Benscoter, B., Page, S., Rein, G., van der Werf, G. R. and Watts A. (2015). Global vulnerability of peatlands to fire and carbon loss, *Nature Geosci*, 8: pp. 11–14.

Varma, A. (2003). The economics of slash and burn: A case study of the 1997–1998 Indonesian forest fires, *Ecological Economics*, 46: pp. 159–171.

Yulianti, N., Hayasaka, H. and Usup, A. (2012). Recent forest and peat fire trends in Indonesia The latest decade by MODIS Hotspot Data, *Global Environmental Research*, 16: pp. 105–116.

12 An Attempt at a Stakeholder Analysis

Joergen Oerstroem Moeller

Visiting Senior Fellow, ISEAS Yusof Ishak Institute, Singapore
Adjunct Professor, Singapore Management University and Copenhagen Business
School Honorary Alumni, University of Copenhagen
jormol@iseas.edu.sg

From an analytical and theoretical perspective, haze offers a string of perplexing and in some cases worrying questions asking for clarification and hopefully solutions; to be useful they need to go beyond a pure economic approach and embrace politics: Haze is a complex phenomenon calling for intersectoral and interdisciplinary analyses.

The victims are the citizens not primarily found in palm oil producing countries, but in adjacent countries. The culprits are the business sector. Governments are under pressure to act, but to a large degree unable to do so. Civil society reacts, but encounters obstacles analogous to those facing the governments. Various groups of stakeholders pursue adversarial objectives.

Fighting haze is a public policy with the purpose of reducing air pollution threatening the health of the population. The inputs come from the public — often an angry public — demanding action here and now and are vocal in their demands. The goals are to force those responsible for haze to change methods of land clearing instead of slash and burn. The instruments can be direct intervention in the form of rules and regulations or economic incentives making alternative methods of land clearing profitable. The effects are easy to monitor in form of a reduction in Pollutant Standard Index (PSI), but difficult to achieve.

Governments run into three main obstacles trying to implement a public policy. The structure of the palm oil industry is not uniform, making it difficult to invent and apply appropriate and effective policy instruments. Slash and burn is the most cost effective land clearing method. The haze originates in one country, but hits citizens in other countries — a transboundary environmental problem.

The business sector may prefer short-term and low cost practices

Palm oil production constitutes an essential part of Indonesia's and Malaysia's economy. It accounts for somewhere between 5 and 10 percent of gross domestic product (GDP), constitutes a sizeable share of exports, offers jobs (for Indonesia more than 3 million people are directly employed by the sector; including jobs generated by the palm oil sector outside the sector itself the figure will be much higher), and through an export tax provides revenue for the government. Both countries look to palm oil as a growth sector over the next decade. The cost for the two countries of haze (health, loss of tourism etc.) may be big, but compared to the significance of the sector for the economy as a whole actually small. No wonder then that both countries are extremely reluctant to introduce measures threatening to kill this goose laying golden eggs.

Measures to reduce haze from palm oil producers would be welcomed, but on the condition that they do not dent on the growth prospect for the sector. Any recommendations or advice out of touch with this reality do not stand much chance of success. The situation is asymmetrical. Indonesia and Malaysia have selected palm oil as a booster for their economies and have decided to tackle external diseconomies along the road. Adjacent countries may be inclined to wish immediate action to reduce external diseconomies — haze — without attaching similar weight to impacts on these two countries' economies. This is the economic circle that has to be squared.

The problem, however, cannot be confined to economics. Other social sciences and ownership structure have to be included. The need for inter-sectoral and interdisciplinary analyses is visible for many developing economics where much advice makes sense from an economic point of view, but disregards societal factors and social fabric. This was one of several flaws in policies adopted by the Bretton Woods institutions pointed out by Joseph Stiglitz.

I studied economics at the University of Copenhagen in the 1960s and one of the professors, who had spent a couple of years as an adviser in a developing country, mentioned such a case. A team from the World Bank, as far as I recall, went to a country and left with the advice that productivity in the agricultural sector and soil fertility should be higher. They recommended the use of more fertilizers. They overlooked though or did not bother to look into contractual links between the landowners and the tenants. The landlords lived in the main cities and left it to tenants to cultivate the land. Tenants were paid a fixed sum of money irrespective of yields. If they wanted higher yields they had to find the money themselves. Use of more fertilizer would hence be paid out of the pocket of the tenants and the higher revenue through higher yields collected by the landlords. No wonder the advice could be classified as completely correct and completely useless.

Presuming rational behaviour — which cannot always be done, however, but necessary for an analysis — a genuine effort to reduce haze will be undertaken only if economically profitable for those who are involved. Persuasion, rules and regulations can reflect this. They can be brought into play, but experience suggests that only with the support of a determined political leadership ready to enforce the rules and an administrative capacity to do so, will this approach be effective. The landowners may in some cases be small farmers and in other cases large corporations based inside and outside Indonesia. The point in this context is that the economic outlook is completely different for these two groups. Politically the small holders and the large corporations leverage influence on the political systems, but not always along similar lines.

The small farmers may have cultivated the land over a long time and in some cases generations not knowing of other ways to clear than slash and burn. It worked well enough in the past — not much air pollution. Several factors have changed the picture. Global climate has turned El Niño into a much more threatening phenomenon — more frequent and stronger impact. Rising population and increasing living standard have stimulated demand for palm oil explaining why production has gone up and a higher production of palm oil worsens the haze. With a lower production level, Southeast Asia might have been able to live with slash and burn, but no longer.

The small farmers may be rooted in traditions. They do not behave differently than their parents and grandparents did, so from their point of view what is the problem? They have limited funds for investment in their plots. Their time horizon may be short and a systematic effort to substitute slash and burn with other methods will be costly in the short term, while possibly profitable in the long run. They cannot afford to wait. It is therefore not difficult to spot why small farmers are holding back, showing reluctance about investing in a changed production process.

The outlook of larger companies differs. They will normally have the funds to invest and even if initial investment may reduce profits it could be profitable in the longer term. They would enhance their reputation as good corporate citizens by acting in accordance with emerging ethical attitudes in nations and communities where they operate.

The snag is that their reputation in Southeast Asia as corporate citizens may not be important as they sell most of their products globally. The major markets are found outside the region — India, the European Union (EU),[1] and China. The rest of the world finds global warming a major issue and companies not intercepting this suffer in reputation, but the same is not the case for haze, which is a regional and not global phenomenon.

Knowing the strong feelings the haze provokes in Southeast Asian countries, companies might find it attractive and profitable to sit back and wait for governmental funds or subsidies to finance all or parts of the cost of reducing haze. Governments are sensitive to people (the voters) so it is possible that governments may blink first and offer to 'help' financially to change production processes. We do not know whether this policy is on the companies' agenda and writing about such things you should be careful to maintain a balanced approach, but we do know that 85 percent of global output comes from Indonesia and Malaysia. Companies have a fairly large room of maneuver in absence of international competition.

---

[1] The European demand discloses a depressing picture. Demand for palm oil has increased substantially over recent years with approximately half of the consumption being used as fuel for vehicles replacing fossil in an effort to reduce greenhouse gas emissions. Unfortunately, research indicates that deforestation taken into account leads to the result that total greenhouse gas emission after such a shift goes up — not down. It illustrates the necessity of broadening the analysis from visible and tangible figures to life cycle and product cycle analysis.

Politically, they have links to the governments and political parties where a parallel can be drawn to the political leverage of the financial sector in the US and the phrase 'too big to fall'.

The most cost effective land clearing method

The right mix of economic incentives combined with rules and regulations might solve the problem by making slash-and-burn unprofitable for the producer — small farmers, large plantations, or multinational companies.

The first step is to realize that in the short run slash and burn probably is economically preferable to other solutions. Or in other words a strict application of market economics will not remove the incentive to continue slash and burn. Short-term market forces must be distorted by subsidies or taxes/levies to change the economic result for investments in other methods to clear the land.

The second step is to ask who should pay. The producers will be unwilling to do so unless they can hike the price to cover their costs and they may not be willing to do so. From a theoretical point of view, it could be argued that the concentration of production makes such a policy possible, but realities tell a different story. The governments may or may not be ready, but do we talk about governments in producing countries or governments in adjacent countries or governments in countries using palm oil? An application of PPP (Polluter-Pays Principle) looks attractive, but as the polluters are consumers all over the world this is difficult to implement and difficult to explain referring to pollution in another geographical areas than where the ultimate consumer is living.

Even if willingness to pay in form of levies (PPP) was forthcoming it might not be attractive for the producers and their home countries. It would mean a higher price for a number of consumer goods using palm oil, putting downward pressure on demand and hence jeopardizing efforts to stimulate palm oil production aimed at boosting the domestic economy. Companies using palm oil in consumer products might start to look for an alternative. Economics tells that a higher price is conducive to research for substitutes.

Welfare economics and game theory have tried to work out a 'theorem' for how bargaining among interested parties can lead to a Pareto optimal solution[2] in case of diseconomies. The basic proposition often applied in public policies is that liability should in principle be assigned to the economic operator for whom avoiding the costs associated with the problem are the lowest. The fundamental thinking is straightforward looking for the lowest costs. There is, however, several obstacles for using this theorem in the haze situation. It is assumed that transparency is high, that costs can be calculated correctly, and that economic operators are able to negotiate with each other on an equal footing plus a public authority willing and capable to providing a framework for such policies. These are assumptions that look unrealistic in the haze situation.

On top of that comes the international aspect not included in the theorem. Producers, consumers and governments are not found in one single country, which undermines the assumption of rational behaviour.

Pollution originating in one country, but mainly deteriorating the environment in adjacent countries

Economic globalization implies that policies pursued by one country affect other countries and primarily adjacent countries. This is why the world has elaborated fairly comprehensive machinery for coordination and consultation of economic policies. Globally, this falls under the Bretton Woods institutions — the World Bank, the International Monetary Fund, and the World Trade Organization. Regionally, similar organizations have emerged.

They work — more or less. Many of them serve more as a forum for consultation than coordination. And that sets limits for how much they can achieve. Countries hit and hit hard by events and policies in adjacent countries do not have much say in policymaking there.

As a classic example of transboundary environmental problem haze is comparable to the location of several nuclear power stations in Europe. Some European countries decided about 40 years ago that they did not wish nuclear power fearing an accident that would lay waste a part of the country. Disregarding these worries, neighboring countries built nuclear

---

[2] A state of allocation of resources in which it is impossible to make any one individual better off without making at least one individual worse off.

power stations just across the border where a river or the sea offered suitable provisions for cooling.

This is a question of sovereignty. The countries ruling out nuclear power exercised their sovereignty, but were nevertheless exposed to risks by adjacent countries' decision without any concern that it could be close to the capital of the neighboring country.

The EU has since introduced pooling of sovereignty with a large number of decisions taken by the EU and not by individual nation-states. A barrier for nuclear power stations threatening a neighboring country rejecting this energy source would have been erected.

This can be generalized to cover air pollution and haze. As long as international cooperation does not interfere with individual member states' prerogative and exclusive right to decide, even vital interests of adjacent countries can be threatened without much leverage for them to prevent it. Negotiations can be initiated. Worries can be voiced. In a worst case scenario, actual threats — sanctions or even armed conflict — may be exchanged, but to no avail if each country insists on exercising its sovereignty.

In Southeast Asia, Association of Southeast Asian Nations (ASEAN) has developed into an organization with many virtues bringing benefits of various kinds to the member states, but it is still far away from pooling of sovereignty. There are arguments for and against taking such a step in this part of the world. In this context suffice to state that as long as this is the case, member states find it difficult to force another member state to adopt policies it does not want to adopt.

According to Bloomberg Singapore, under the Transboundary Haze Pollution Act of 2014, six suppliers of Indonesia's Asia Pulp and Paper Group were ordered to provide information on steps they are taking to prevent fires on their land. Indonesian President Joko Widodo backed Singapore's plans to wield heftier fines against overseas polluters as long as sovereignty is respected, before he took office in 2014.

Concluding remarks

A traditional stakeholder analysis bringing in business, government, and society to look at adversarial or mutually beneficial goals for a public policy is difficult to undertake as haze is a transboundary problem. To this must be added its asymmetrical nature with strong economic interests involved for the two producing countries, Indonesia and Malaysia, compared to negative effects of tangible nature, but smaller relative to GDP for non-producing countries.

Civil society in producing countries and non-producing countries can play a role. Governments can introduce various rules to control slash and burn and step in to douse the fires where possible. These are useful steps.

It remains, however, difficult to escape the observation that as long as producers find it profitable to slash and burn they will continue to do so albeit gradually shifting to other and costlier methods. The main effort for governments thus lies with introducing economic incentives and technology plus logistics that can change the economic calculation from favoring slash and burn to promote alternative methods.

13 The Gross Underestimation of the Costs of Haze[1]

Ng Yew-Kwang

Winsemius Professor in Economics, Division of Economics,
Nanyang Technological University, Singapore
ykng@ntu.edu.sg

The transboundary haze is the most important factor affecting the otherwise very clean environment of Singapore. It is a very important factor affecting whether people will stay or come to Singapore or leave. In Chinese cities, it has recently been reported that whenever the smog level was high, many more people obtained information regarding emigration. In Singapore, with the haze situation getting so bad around September/October 2015, similar considerations may prevail. I myself for one had considered Singapore a possible location for long-term retirement living. But the persistent haze in 2015 made me have a second thought. The haze situation in 2013, though also sharp at times, did not persist for as long as the one in 2015. However, the more important reason to avoid the haze is the health (and hence welfare) of people who will be here with or without the haze.

In fact, people in Indonesia, especially in Sumatra, suffer more from haze caused by the burning of forest for planting than people in neighboring countries including Singapore. Thus it is not very helpful to model the problem as a game-theoretic situation involving an external cost produced by country A on countries B and C. It is more a problem of governance failure. However, due to the international and even global effects, some

---

[1] Some of the views in this paper appeared in articles published in *Business Times, Lianhe Zaobao*, and *Straits Times* over 2013/14.

international collaboration may help solve the governance failure, as discussed below.

The gross underestimation of the costs of haze

Many analysts discussing the costs of the haze (to Singapore for simplicity here) focus on the economic costs or the effects on GDP only, such as the costs of masks, increased hospitalization, and tourism revenues lost. Though these are not completely irrelevant, they are NOT the main costs. To discuss the true costs of the haze, let us first focus on those imposed on people in Singapore as individuals, before considering wider costs like global warming on the whole world, and Singaporeans as producers.

For people in Singapore as individuals, the main costs are the negative health effects of the haze and the secondary costs are the costs associated with preventive measures like wearing masks and avoiding outdoor activities. Adding these two types (main and secondary) of costs together does not involve double counting if the main negative health costs are measured/estimated with preventive measures. Except perhaps for people with very low incomes, the costs associated with preventive measures are not mainly the tangible ones like the costs of masks. Rather, they are the nuisance of having to wear masks (which could be hundreds of times more than the cost of a mask for many people) and the opportunity benefits of activities forgone, like not being able to walk and exercise outdoor. Similarly, the health costs are mainly not the costs of higher medical bills but the costs of actually lower health levels (plus the costs of larger medical bills).

Clearly, the true costs of haze in the form of prevention costs and negative health costs could be very large even if there is no loss of tourist revenues and GDP is not affected. In fact, GDP may increase despite the true costs being very high. For example, before the 2015 haze, I walked to and from work. During the haze, I took taxis. Similarly, more preventive and treatment measures involving more goods and services may be incurred and provided. The GDP could increase but people are certainly made worse off.

Though the reduction in tourism revenues is a loss to Singapore, treating the whole amount of revenue decline as a loss is questionable economics, as it ignores the costs involved in earning those revenues, ignores possible intertemporal substitution (tourists not coming to Singapore in September 2015 may come in May 2016 instead; this does not preclude the likely decrease in overall tourism revenues by a smaller amount), and ignores long-run alternative options other than tourism. In the very short run, the facilities catering to tourism are largely fixed, a fluctuation in demand mainly manifests in gains or losses in revenue. However, in the longer run, if the situation persists, resources used in catering for tourism may be transferred to other industries. To see this point more clearly, consider a hypothetical and dramatized case. Singapore was a tourism-focused economy with 80 percent of its GDP earned from tourism. However, increasing haze from Indonesia gradually reduced its tourism from 80 percent to 10 percent of GDP only over three decades. We cannot then conclude that the haze costed Singapore 70 percent of its GDP, as the resources involved would have gradually been transferred to other industries. On the other hand, even if 100 percent of its resources remain employed, we cannot say that there were no losses either. If Singapore really had comparative advantage in tourism, the transfer to other industries forced by tourism decline caused by haze would likely have reduced its GDP. The true losses associated with the decline in tourism would be something of a triangle, not a larger rectangle. However, the losses of negative health effects should not be forgotten.

Since we likely have over-estimation in factors like tourism losses and under-estimation like health effects, could we accept the usual estimates as roughly OK? No! In my view, the negative health effects, if properly estimated, would trump all other over-estimation by many many times. Before a proper analysis of the true costs of haze has been undertaken, we may do a very rough guestimate.

Economists typically use 'willingness to pay' or 'needed amount of compensation' (willingness to accept), or an average of the two, to estimate the benefits or costs of an event. To try to gauge the costs of the haze crisis, I ask myself what percentage of my annual salary I would be willing to forgo to make the problem (defined as the haze level over the three months period in 2015) go away. My answer is: Not less than 20 percent. But what if the skies continue to remain hazy? How much compensation, in terms of percentage of my annual pay, will make me indifferent to the situation without the problem? My answer: Not less than 40 percent. The average of the 20 percent and 40 percent is 30 percent. This means that a rough estimate of costs of haze at an intensity and duration of 2015 to me is no less than 30 percent of my annual income. Let us conservatively just use the lower willing-to-pay figure of 20 percent. Suppose your annual salary is 100,000 dollars, 20 percent of which is 20,000 dollars. This is equivalent to 80 percent of your quarterly salary of 25,000 dollars. The costs of haze could be as high as 80 percent of incomes over the affected period. [It should be noted that the willingness to pay might be affected by the ability of payment from accumulated wealth or financed by future incomes. These factors are not explicitly considered for simplicity.]

However, my willingness to pay to avoid the haze (as a percentage of earned income) may be much higher than that of the average Singaporean. Let us suppose, for the sake of argument, that the average Singaporean's 'willingness to pay' to put up with the haze is only a fifth of mine. This works out to be 16 percent of his or her earned incomes during the period of haze. The median monthly household income from work is 8,666 Singapore dollars in 2015, according to the Department of Statistics. Sixteen percent of that figure is 1,386 Singapore dollars, or about 46 Singapore dollars per day. That alone (before counting other costs discussed below) is around 46 times the a-dollar-a-day damages per person estimated by some analysts (5 million Singapore dollars per day for the whole of Singapore). Think about this. If it were a-dollar-a-day problem, could it be such an important issue? Come on!

However, the gold medal for underestimating the costs of haze does not go to any of these analysts. Rather, I propose to award it to the Singapore's Transboundary Haze Law of 2014. Under this law, fines are up to 100,000 Singapore dollars a day, capped at a total of 2 million Singapore dollars, for causing unhealthy haze, defined as a Pollutant Standards Index value of 101 or greater for 24 hours or more. Some people may think that 100,000 Singapore dollars a day is a big penalty. However, since the haze affects all people in Singapore, that sum is less than two cents a day per person. Two cents per day! The whole period of haze for three months is less important than a cup of coffee! The gold medal is well deserved! This is certainly well less than 1 percent of any reasonable estimate of the costs of haze at an unhealthy level, and is less than 0.04 percent of my 46

Singapore dollars guestimate above. There are good grounds for increasing the maximum fines by at least a hundred, if not a thousand, times. Though the then Environment and Water Resources minister Vivian Balakrishnan promised to stiffen the penalties further if necessary, it would be more effective to have higher figures right away to serve as a strong signal.

In addition, convicting the culprits for this case of transboundary haze is likely to be very difficult and costly. As the probabilities of conviction are low, the expected fines (fines time probabilities) are much lower than the maximum fines provided. This makes it much more important to provide for more adequate maximum fines. The provision for insufficient fines is probably partly affected by the tendency to focus on the more tangible costs only, ignoring the true health costs mentioned above.

In addition, going beyond traditional economics, there are good reasons to suggest that most people (myself included) tend to underestimate the importance (and hence also the amount of money they should be willing to pay) of unhealthy things with effects likely to occur in the future with uncertainties. This suggests that a government concerned with the long-term welfare of the people should undertake even more effort and spend more money to avoid such unhealthy things like haze, than indicated by people's willingness to pay.

Secondly, happiness studies suggest that, beyond the levels of biological survival and some low level of comfort, higher consumption does not significantly increase happiness, especially at the social level (e.g., Diener *et al.*, 2010; Ng, 2015). This again suggests that we should do more and be willing to spend more to get rid of such definitely health-reducing and hence happiness-reducing things like haze. We should definitely not be confined to the narrow cost-benefit calculus taking account of only tangible economic effects like tourism and GDP. Health and happiness are far more important. This is especially so for Singapore now as we are way beyond the level decades ago when GDP was more important.

The discussion above, including the guestimate of the 46 Singapore dollars a day per person cost of haze, focuses on the costs faced by Singaporeans as individuals. What about Singaporeans as producers? Costs like losses to agricultural crops, fire-fighting costs, etc. apply mainly to Indonesia itself and to some extent, to Malaysia. These costs are not significant for Singapore. More importantly, if we have done a proper cost analysis based on willingness to pay/accept, and if we have covered all (or a representative sample of) individuals in Singapore, no additional costs to Singaporeans as producers should be included. This is so because all producers are also ultimately individuals (companies are owned by individuals). For example, consider the losses of work productivity. If these are properly perceived, they should already be reflected in the willing to pay/accept figures. On the other hand, it is also quite possible that, when considering their willingness to pay/accept the haze, most individuals may only take into account the perceived direct health costs, but not the more indirect costs of their productivity reduction in the future, and the possible reduction in the share prices of companies they hold. If so, separate estimates of these costs on the producer side may have to be added. In addition, if we consider people beyond Singapore, there may be losses to foreign owners of companies in Singapore which are not reflected in the willingness to pay from local Singaporeans only. Also, the inability to visit Singapore on their preferred period due to haze may also impose costs on these potential tourists or businesspersons. However, the amounts involved are unlikely to be large, except for people who have important business in Singapore over the relevant period.

If we look beyond Singapore, the most important costs of haze apart from the main negative health effects are the negative effects on global warming, including the loss of forests and the burning with its resulting pollution and emission of greenhouse gases. These are largely global effects. If a long-term view is taken as should be, and if a threat to global survival is involved (Ng, 2016), the costs of the forest fires involved could be much higher than the already huge health costs discussed above. However, the estimation of these global costs is beyond the scope of this chapter.

Towards a solution of the problem

As mentioned above, the problem of forest fires and the resulting haze is a governance failure, the failure of the Indonesian government at various levels to protect its own people and the people in neighboring countries and the whole world from the inefficient external costs of forest burning.

This is inefficient because the lower costs of clearing the forest by burning are much more than offset by the external costs imposed on others. As a governance failure, the solution probably has to rely largely on the Indonesian polity. However, since the problem also involves transboundary haze and global climate change, some international effort may help towards the solution.

Malaysia and Singapore have involved with discussion with Indonesia. In September 2014, Indonesia also ratified the ASEAN haze agreement. Reports up to May 2015 also suggest that Indonesia has been more active in trying to prevent forest fires. However, regional agreement and effort alone may not be sufficient, as they tackle only the regional external costs of haze but not the global external costs of exacerbating climate change and the external benefits of preserving forests. The preservation of forests produces environmental external benefits good for the whole world, not to mention the additional concern for the threatened species like orangutan. In addition, according to the World Health Organization, the consumption of palm oil should be reduced. Thus, the burning of forest for palm plantation has several serious adverse effects.

Malaysia is to be commended for agreeing to preserve at least half of its forest, but this may not be enough. Unless the benefits of development exceed the sum total of the costs, including external costs like air pollution, of conversion and the lost benefits of forest preservation, development is undesirable. However, these costs and benefits are not confined to the country of deforestation, but also affect neighboring countries and the whole world. The country concerned does not have sufficient incentives. An international or even worldwide approach may be needed.

Malaysia and Singapore should consider the option of joining forces and appealing to the United Nations (UN). I do not mean to accuse Indonesia or sue for compensation. Rather, from the perspective of economics, since the present situation is very inefficient, some improved arrangement should be possible such that every country could be made better off. The world should pay Malaysia and Indonesia for preserving their forests; the polluters should also be penalized for the expected damages imposed on others. The UN's REDD (Reducing Emissions from Deforestation and Forest Degradation) idea (and REDD+) of giving developing countries the financial incentive to keep their forests and other carbon stores is desirable, but more are needed.

With the advancement of science and technology, economic growth and globalization, terrorism and environmental disruption, the world has become much more integrated. The 200-odd countries in the world cannot just each look after its own backyard. A country needs a national government; the 200-odd countries need a world government. True, we have the United Nations. However, the UN is relatively much less powerful than a national government. The budget of a national government is typically in the order of 30–40 percent of the GNP, but the UN budget is less than 0.01 percent of the global gross national product (GNP), a relative difference of many thousand times! In addition, the UN has much less power of coercion than a national government. The UN needs to be empowered to do its job properly.

However, before we have successfully tackled the question of democratic representation and control, a powerful government could also be tyrannical. Before that long-term stage of a democratic and responsible world government, we may first aim for an intermediate stage of empowering the UN by increasing membership contributions and cooperation with the WTO (World Trade Organization), using trade sanction against gross violators of environmental quality, as argued in Ng and Liu (2003).

However, most economists are typically in favor of free trade and reluctant to use the sanction of WTO to achieve other objectives. Within a country, all individuals also find individual freedom very important, but that does not preclude the need to imprison certain criminals who impose great harms on others! To achieve the greater objective of environmental protection, WTO should cooperate with the UN to use trade and investment sanctions against gross violators of environmental quality such as the burning of forests.

Many environmental scientists believe that, due to the existence of tipping points and cascading effects, we may only have a few decades left to prevent catastrophic destruction of the living environment of the whole world. If Malaysia and Singapore could successfully approach the UN with a sensible proposal to empower the UN (hence likely to be viewed favorably) and to help solve regional and global environmental problems, the recent haze could be a blessing in disguise!

References

Diener, E., Kahneman, D. and Helliwell, J. (2010). *International Differences in Well-Being*, Oxford University Press: Oxford.

Ng, Y-K. (2015). *Happinessism*, Taiwan: Wunan Press. (In Chinese.)

Ng, Y-K. (2016). The importance of global extinction in climate change policy, *Global Policy*, in print.

Ng, Y-K. and Liu, P-T. (2003). Global environmental protection — Solving the international public-good problem by empowering the United Nations through cooperation with WTO, *International Journal of Global Environmental Issues*, 3(4): 409–417.

14 How Singapore Can Help Stop Haze: The Palm Oil Factor

PM Haze

Background to the haze

The Southeast Asian haze is a decades old occurrence, and has been recorded in Singapore since 1972.[1] The haze is caused by large-scale fires in Indonesia and Malaysia, where millions of hectares may be burnt in a year. The recent crisis between June and October 2015 resulted in a burnt area of 2.6 million hectares.[2]

The haze is a toxic mix of harmful gases such as carbon monoxide, ammonia, cyanide and formaldehyde,[3] as well as microscopic particles coated with carcinogens such as polycyclic aromatic hydrocarbons.[4]

Health effects of the haze are caused when there is an irritant effect of fine dust particles on the nose, throat, airways, skin and eyes. Individuals with medical conditions like asthma, chronic lung disease and skin conditions can experience severe symptoms.

The haze not only has severe impacts on human health but also has an impact on the environment, society, and economy. The 2015 haze cost Singapore an estimated 700 million Singapore dollars.[5]

---

[1] Lee, M. K. (2015). Haze in Singapore: A problem dating back 40 years. *The Straits Times*.

[2] World Bank. (2016). The cost of fire: An economic analysis of Indonesia's 2015 fire crisis. *Indonesia Sustainable Landscapes Knowledge Note*, 1.

[3] Dayne, S. (2015). Don't inhale: Scientists look at what the Indonesian haze is made of. *Forest News*, CIFOR.

[4] Pavagadhi, S., Betha, R., Venkatesan, S., Balasubramanian, R. and Hande, M. P. (2013). Physicochemical and toxicological characteristics of urban aerosols during a recent Indonesian biomass burning episode. *Environ. Sci. Pollut. Res.*, 20: 2569–2578.

[5] Barratt, O. (2016). Haze episode cost Singapore estimated S$700m last year: Masagos. *Channel NewsAsia*.

Causes of the haze

Although historically, fires have occurred in the rainforests of Indonesia during periods of extreme drought,[6] in recent decades, the fires and haze have increased in frequency and severity.[7]

Mainstream media has tended to focus on use of fire to clear land or 'slash-and-burn' as the cause of haze.[8] Yet, swidden agriculture has been used traditionally by indigenous farmers on a small scale for hundreds of years. They used fire to prepare rotating swidden fields of about 1.5 hectare each in a system also known as shifting cultivation.[9]

What were the key changes that occurred in recent decades that led to large-scale fires and haze?

It is combination of larger scale and more frequent use of fire to clear land, claim tenure or as a weapon; emergence of fire-prone landscapes due to deforestation and peat drainage; and the inability or unwillingness to stop fires early due to poor fire-fighting capacity and lack of governance.

From the 1980s, road construction by logging companies and transmigration projects opened up land, making access easy for migrants. Migrant farmers used slash-and-burn techniques followed by tree planting to secure tenure over land.[10] With the development of industrial-scale plantations, fires have sometimes been used in an uncontrolled manner to clear vegetation on a large scale. Farmers use fire for the clearing of new land because it is much cheaper than the mechanical alternative where it can cost up to 200 US dollars for equipment and chemicals to clear one hectare of land, compared to 5 US dollars using fire.[11] Starting a fire can also clear

---

[6] Cochrane, M. A. (2003). Fire science for rainforests. *Nature*, 421: 913–919.

[7] Field, R. D., van der Werf, G. R. and Shen, S. S. P. (2009). Human amplification of drought-induced biomass burning in Indonesia since 1960. *Nat. Geosci.*, 2: 185–188.

[8] Forsyth, T. (2014). Public concerns about transboundary haze: A comparison of Indonesia, Singapore and Malaysia. *Glob. Environ. Chang.*, 25: 76–86.

[9] Goenner, C. (2000). Causes and impacts of forest fires: A case study from East Kalimantan, Indonesia. *International Forest Fire News*.

[10] Palm, C. A., Vosti, S. A., Sanchez, P. A., Tomich, T. P. & Kasyoki, J. (2005). Alternatives to slash and burn. in Palm, C. A., Vosti, S. A., Sanchez, P. A. and Ericksen, P. J. (eds.), *Slash-and-Burn Agriculture*, Columbia University Press, pp. 3–37.

[11] Varkkey, H. (2013). Patronage politics, plantation fires and transboundary haze. *Environ. Hazards*, 12: 200–217.

the plant materials, which kills pests, fertilises the soil and neutralises acidity at a much lower cost.

Besides companies, some also blame use of fire on poor small-scale farmers expanding their farmland, and rogue operators involved in illegal clearance to acquire more land.[12] Fires started can easily go out of control, especially when used by medium-sized companies which usually lack fire management resources or rogue operators that may not stay to monitor the fires for fear of getting caught.

Land rights of communities in Indonesia are unclear and often generally unrecognized by the government. These communities do not have the ability to enforce laws and there are often competing claims over a plot of land. The arrival of new, external actors such as transmigrants and plantation operators onto land occupied by local communities increases the likelihood that fire will be used as a weapon.[13]

Peat is a type of soil that is composed mainly of partially decayed plant matter formed over thousands of years in waterlogged areas known as peat swamps. The peatland ecosystem is the most efficient carbon sink on the planet, but it has taken thousands of years for peatlands to develop these deposits. About 14 percent of Indonesia and 8 percent of Malaysia are covered by peat.[14] Like a sponge, peat domes store water during the wet season, reducing the risk of flood, and release water slowly during the dry season, mitigating the risk of drought and fires.

In order to plant conventional crops or for logging, peat swamps are drained, creating fire-prone landscapes which fuel the spread of fires. From 1990 to 2015, half of the peat swamps forest in Peninsular Malaysia, Sumatra and Borneo were cleared.[15]

---

[12] Balch, O. (2015). Indonesia's forest fires: everything you need to know. *The Guardian*.

[13] Colfer, C. J. P. (2002). Ten propositions to explain Kalimantan's fires. in Colfer, C. J. P. and Resosudarmo, I. A. P. (eds.), *Which Way Forward? People, Forests, and Policymaking*, Resources for the Future, pp. 309–324.

[14] Joosten, H. (2009). The global peatland CO2 picture: Peatland status and drainage related emissions in all countries of the world, *Wetl. Int.*, 36.

[15] Miettinen, J., Shi, C. and Liew, S. C. (2016). Land cover distribution in the peatlands of Peninsular Malaysia , Sumatra and Borneo in 2015 with changes since 1990, *Glob. Ecol. Conserv.*, 6: 67–78.

Peatlands are carbon rich and, while waterlogged, are not prone to fire. But dried, carbon rich peat soil can catch fire easily. During the dry season, a single discarded cigarette can cause a fire in drained peatland.[16]

Peat fires can smoulder underground for days, or even months.[17] Putting out peat fire requires huge volume of water to soak through the soil to extinguish the fire below.

In addition, continued deforestation results in less trees, causing a drier local climate, which in turn reduces the soil moisture level, making the area more fire prone.[18]

Despite possessing greater financial capacity, even multinational agribusinesses have had trouble protecting their crops from fire,[19] let alone farmers and smallholders who often lack necessary firefighting capacity. Limitations in firefighting capacity on the ground include insufficient manpower and training, insufficient or poor equipment and lack of access to water. Early detection and suppression of fires may also be hampered by poor monitoring for example due to access to data such as satellite hotspots, inaccuracy of satellite hotspot data and lack of equipment such as drones, cameras and fire lookout towers.

Land tenure in Indonesia is frequently unclear and disputed. This can lead to situations in which culpability and responsibility are difficult to establish and in which individuals or companies are rarely held accountable for fires.[20] Illegal land clearing, land grabbing and land transactions with illegal companies are still rampant. For example, at the village level, elites may sell land to illegal companies on a conservation area gazetted by central government or a district government might extend concessions to companies.[21] Land concessions might also overlap, resulting in conflicts and disputes.

---

[16] Goenner, C. (2000). Causes and impacts of forest fires: A case study from East Kalimantan, Indonesia. *International Forest Fire News*.

[17] Turetsky, M. R. *et al.* (2015). Global vulnerability of peatlands to fire and carbon loss, *Nat. Geosci*, 8: 11–14.

[18] Uhl, C. and Kauffman, J. B. (1990). Deforestation, fire susceptibility, and potential tree responses to fire in the Eastern Amazon, *Ecology*, 71: 437–449.

[19] Hutan Kita Institute — HaKI *et al.* (2016). 2015 fires burned 26% of APP's plantations in South Sumatra, raising questions about fiber supply for new OKI mill, 1–5.

[20] Gaveau, D. L. A. *et al.* (2017). Overlapping land claims limit the use of satellites to monitor no-deforestation commitments and no-burning compliance, 10: 257–264.

[21] Purnomo, H. *et al.* (2017). Fire economy and actor network of forest and land fires in Indonesia. *For. Policy Econ.*, 78: 21–31.

Why palm oil is key to stopping the haze

Palm oil is highly versatile, and is used in cooking oil and as shortening in food, for the manufacture of greases, lubricants, candles, and as feedstock for the production of biodiesel. Palm oil derivatives are used in cosmetics, pharmaceuticals, bactericides, and water-treatment products.

Global demand for palm oil has resulted in an explosion in palm oil production. Since the 1980s, palm oil production has increased tenfold, and is now the most widely used vegetable oil in the world. The first commercial palm oil plantation was established in Selangor Malaysia in 1917 and today Malaysia and Indonesia together produce about 85 percent of global palm oil.[22]

Much of this expansion has come at the expense of tropical rainforests and peat swamp forests. Between 2001 and 2010, half the deforestation in Indonesia and a third of deforestation in Malaysia could be attributed to expansion of oil palm plantations.[23]

Palm oil companies also make up a significant proportion of companies tried in court for causing fires. For example, in 2014, palm oil company PT Kallista Alam was found guilty of illegally clearing and burning forest in the Tripa peatland of Sumatra, Indonesia.[24]

The palm oil industry has a complex supply chain (see Figure 1) involving multiple changes of corporate ownership as the palm oil passes from growers to mills then on to traders, processors, manufacturers and finally to the food we dine on or products that we buy.

Within this complex supply chain, numerous mid-level palm oil operations remain obscure even as they commit blatant abuses like clearing protected forests. Illegal oil palm plantations in Tesso Nilo National Park have been linked with widespread deforestation and fires in the national park.[25]

---

[22] FAO. (2017). FAOSTAT database collections.

[23] Henders, S., Persson, U. M. and Kastner, T. (2015). Trading forests: Land-use change and carbon emissions embodied in production and exports of forest-risk commodities, *Environ. Res. Lett.*, 10.

[24] Butler, R. A. (2014). In precedent-setting case, palm oil company fined $30M for destroying orangutan forest. *Mongabay*.

[25] Minnemeyer, S., Samadhi, T., Anderson, J. and Gong, M. (2015). Forest fires blaze in Indonesia's Tesso Nilo National Park. *World Resources Institute*.

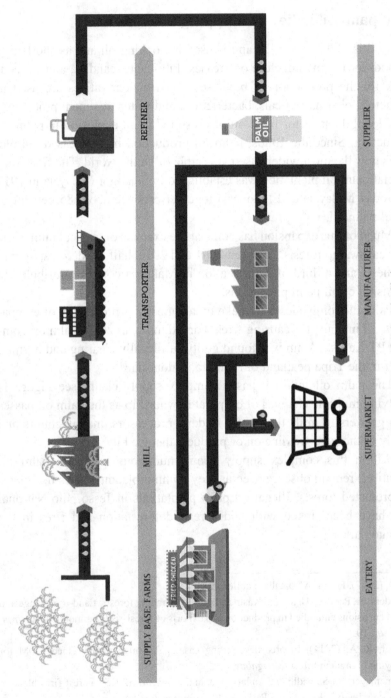

Figure 1. Palm Oil Supply Chain

Due to the difficulties in traceability, mills may not be aware exactly where their oil palm fruits originate. In this way, several large palm oil producers that made No Deforestation pledges have been found to be accepting palm oil fruits grown illegally in Tesso Nilo National Park.[26]

Even with greater demand for sustainable palm oil from consumers, non-governmental organizations (NGOs) and companies in some countries, some producers choose to continue with their destructive practices by selling to less discerning markets. Sawit Sumbermas Sarana, which has been accused of clearing peatlands and forests, lost two buyers with strong No Deforestation, No Peat and No Exploitation (NDPE) policies.[27] But it then found buyers less concerned with sustainability and now sells mostly to companies producing cooking oil for Indonesia's market.

Singapore's hand in the palm oil industry

When compared with global production of palm oil, the amount of palm oil consumed in Singapore may seem insignificant. In 2016, Singapore's net imports of palm oil and palm kernel oil amounted to 0.4 million tonnes, out of global production of 69.2 million tonnes.[28]

The per capita consumption of 0.082 tonnes/person is about 10 times higher than the global average of 0.0085 tonnes/person.[29] This is because one of the largest biodiesel plants in the world is located in Singapore, and uses palm oil, palm waste and waste animal fat as feedstock.[30]

Palm oil is also widely used as cooking oil by eateries in Singapore. According to a PM Haze survey, 46 out of 75 eateries surveyed use palm oil used pure palm oil, or a blend containing palm oil for cooking.

Many of these eateries have a strong regional presence as well. Old Chang Kee, for example, has outlets in Australia, Indonesia and Malaysia,

---

[26] Eyes on the Forest. (2016). No one is safe.

[27] MacIsaac, T. (2016). Consumer pressure to ditch deforestation begins to reach Indonesia's oil palm plantation giants. *Eco-Business*.

[28] Index Mundi. (2017). Indonesia palm oil domestic consumption by year. Retrieved http://www.indexmundi.com/agriculture/?country=id&commodity=palm-oil&graph=domestic-consumption, accessed 15 September 2017.

[29] FAO. (2017). FAOSTAT database collections.

[30] Eco-Business. (2011). Biofuels development: Singapore powering ahead, but gaps remain. *Eco-Business*.

Table 1.   Palm Oil Growers, Traders and Processors with Presence in Singapore

Name	Presence in SG	Listed on SGX?
Wilmar International	HQ	Yes
Olam International Ltd	HQ	Yes
Golden Agri Resources Ltd	HQ	Yes
First Resources Ltd	HQ	Yes
Bumitama Agri Ltd	HQ	Yes
Indofood Agri Resources	HQ	Yes
Kencana Agri Ltd	HQ	Yes
Mewah International Inc	HQ	Yes
Global Palm Resources Holdings Ltd	NA	Yes
Musim Mas Group PT	HQ	N.A.
Cargill Inc	Asia-Pacific hub	N.A.

while BreadTalk Group has operations in 17 countries including China and Indonesia. This provides an opportunity for corporate practices initiated in Singapore to influence larger markets in the region that till date have not shown strong consumer demand for sustainable palm oil.

As a regional commerce and finance hub, some of the largest palm oil growers and traders in the world are headquartered in Singapore (see Table 1). These companies can hold enormous leverage over suppliers who do not comply with best practices.

Financial institutions provide funding to the palm oil industry either as creditors (i.e., providing loans) or investors.

Singapore listed banks DBS, OCBC and UOB are major financiers of the palm oil industry in the region. Therefore they have a key role in influencing practices on the ground such as clearing of land by fire and cultivation on peat.

Although the three banks do not publish details of their loans to palm oil and forest risk sectors, reports published by NGOs and research houses indicate clearly the level of lending of the three banks to those sectors.

A study on loans provided to 16 major palm oil companies found that OCBC, DBS and UOB were among the 15 top providers of loans to these companies.[31]

---

[31] Climate Reaction Research. (2017). Banks finance more palm oil than investors: Investors face indirect exposure.

Another study of 180 forest risk companies in palm oil, pulp & paper, timber and rubber industries revealed that OCBC and DBS were among their top 10 financiers from 2010 to 2016.[32]

Assessment of palm oil alternatives

Palm oil is the most productive vegetable oil crop, producing more oil per hectare than all other oil crops (see Figure 2).

A life cycle analysis of various oils found that palm oil has "medium" impact on greenhouse gas emissions, while rapeseed and sunflower oil have "low impact".[33] However, the study also showed that peat soil decay and methane emissions from palm oil mill effluent are major contributors to palm oil's greenhouse gas emissions — both of which can be avoided by adopting sustainable practices.

Also, a boycott of palm oil would simply drive palm oil producers to sell to less discerning markets. Thus not providing any incentive for existing growers to improve their practices.

Dietary guidelines in many countries discourage the consumption of food with high saturated fat, such as palm oil. Singapore's Health Promotion Board for example provides the healthier choice symbol to vegetable oils with lower saturated fat content.[34] However, while saturated

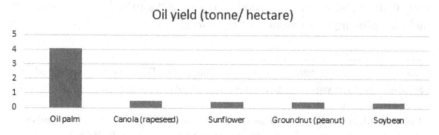

Figure 2. Oil Yields of the Major Vegetable Crops[35]

[32] Rainforest Action Network. (2017). Tuk INDONESIA & Profundo. *Forests & Finance.*
[33] Schmidt, J. H. (2015). Life cycle assessment of five vegetable oils. *J. Clean. Prod.,* 87: 130–138.
[34] Singapore Health Promotion Board. (2016). Healthier Choice Symbol nutrient guidelines, 13.
[35] Murphy, D. (2009). Global oil yields: Have we got it seriously wrong? *AOCS.*

fat intake has been linked with higher levels of low-density lipoprotein (LDL) cholesterol,[36] which is in turn linked to higher risk of heart disease and mortality, a recent study across 18 countries has shown that higher saturated fat intake is associated with lower cardiovascular disease and mortality.[37] The conflicting results may be due to the overall diet intake as many studies that linked higher saturated fat intake with poor health were done in Western countries.[38] Overall, there is lack of clear evidence to support a shift from palm oil to other types of vegetable oil.

Assessment of consumption reduction as a solution

The haze issue is mainly due to expansion of plantations, which in turn is mainly due to continuously growing demand for palm oil.

To reduce pressure to expand palm oil planted areas, there can be either increase in yield or reduction in demand.

However, as mentioned earlier, switching to other oil crops would lead to more land area needed to produce the same amount of oil. One solution should then be to reduce the overall consumption of all vegetable oils, not limited to palm oil.

A study showed that under a scenario in which both developed and developing countries converged to a "healthy" level of vegetable oil consumption, the demand for land would be reduced by 70 percent.[39] The food using most oil is fried food. Possible ways to reduce consumption include reducing consumption of fried food.

---

[36] Sun, Y. *et al.* (2015). Palm oil consumption increases LDL cholesterol compared with vegetable oils low in saturated fat in a meta-analysis of clinical trials, *J. Nutr.* doi:10.3945/jn.115.210575.Palm

[37] Dehghan, M. *et al.* (2017). Associations of fats and carbohydrate intake with cardiovascular disease and mortality in 18 countries from five continents (PURE): A prospective cohort study, *Lancet*, 6736: 1–13.

[38] Dehghan, M. *et al.* (2017). Associations of fats and carbohydrate intake with cardiovascular disease and mortality in 18 countries from five continents (PURE): A prospective cohort study, *Lancet*, 6736: 1–13.

[39] Koh, L. P. and Lee, T. M. (2012). Sensible consumerism for environmental sustainability, *Biol. Conserv.*, 151: 3–6.

Is haze-free palm oil possible?

Growers can produce "haze-free" palm oil by addressing the causes for fires to start, spread and stop:

- Prevent start of fire — Zero-burning and zero land conflict
- Prevent spread of fire — Avoid creating fire-prone landscapes
- Detect and stop fires early

To prevent the starting of fires, machines or tools should be used to clear land (mechanical clearing). When fire is not used to clear land, the method is called "zero-burning". Companies should ensure their own plantations and those of their suppliers use zero-burning methods while assisting local communities living in and around their concessions to use zero-burning methods. To prevent fraud, land burnt by fire should be restored to its original state instead of converting it to commercial use. To reduce the risk of fire being used as a weapon in land conflicts, companies opening new plantations should respect land rights of local communities. Free, Prior and Informed Consent (FPIC) should be obtained and respected.

To avoid creating fire-prone landscapes, growers should avoid clearing forests and draining peatland, and instead develop new plantations only on non-forest land on mineral soil. For existing plantations on peatland, water levels should be maintained as high as possible. Small dams, or "canal blocks", can be built over canals in order to retain water in peat. To avoid risk of fire, groundwater level should be no lower than 50 centimeters below the ground surface.[40] Effective water management also requires a landscape approach across the entire peat hydrology unit where the water level should be kept much higher within the peat dome compared with surrounding areas.

---

[40] Hayasaka, H., Takahashi, H., Limin, S., Yulianti, N. and Usup, A. (2016). Peat fire occurrence, in Osaki, M. & Tsuji, N. (eds.), *Tropical Peatland Ecosystems* (eds.), Springer, pp. 377–396.

During the dry season, the water level in peatland will continue to fall despite the presence of canal blocks due to evapo-transpiration. To eliminate the risk of fire, the peatland should ideally be restored to its wet condition. Crops suitable for growing in wet conditions may be planted instead. Companies should also conserve forests and peat swamps in and around their concession to serve as buffers and barriers to the spread of fire.

To detect and stop fires early, companies should have sufficient equipment and manpower for detecting and fighting fires within and around their concessions. Companies should also equip local communities in and around their concessions to detect and fight fires.

From the perspective of oil palm growers, adopting haze-free practices usually requires an initial investment, for example to buy machinery to clear land without fire and to identify and set aside forests and peatland.[41] Some growers, especially smallholders, may lack the required knowledge, tools or capital to adopt such practices. Encouraging growers to go haze free requires economic incentives for haze-free practices and disincentives for haze-causing practices. These haze-free and haze-causing practices must have clearly defined standards. Supporting systems have to be in place to communicate these standards to the growers and help them build capacity to meet these standards. Compliance then has to be verified, which also requires transparency and traceability.

Once compliance is verified, access to market and to finance can be used as incentives.

Regulated standards

These standards have a regulatory body that ensures all companies comply with a set of criteria.

Certification standards covered are Roundtable on Sustainable Palm Oil (RSPO) (including RSPO Next) and Rainforest Alliance (RA). They have a global reach and include end product labelling to allow consumers to verify their product.

Roundtable on Sustainable Palm Oil (RSPO) is a non-profit organization established in 2004 that encompasses stakeholders from seven

---

[41] WWF, FMO and CDC. (2012). Profitability and Sustainability in Palm Oil Production.

sectors of the palm oil industry: the oil palm producers, processors or traders, consumer goods manufacturers, retailers, banks/investors, and environmental and social NGOs, to develop and implement global standards for sustainable palm oil.

RSPO Next is a voluntary add-on to the standard RSPO criteria for palm oil producers.

Rainforest Alliance (RA) is a non-profit organization that certifies more than 100 different varieties of crops, including oil palm, based on the Sustainable Agriculture Network (SAN) standard.

Mandatory national standards covered are **Indonesian Sustainable Palm Oil (ISPO)**, launched in 2011 and **Malaysian Sustainable Palm Oil (MSPO)**, launched in 2015. These standards are applicable to all oil palm plantations in their respective countries.

Voluntary initiative covered is **Palm Oil Innovation Group (POIG)** which is a multi-stakeholder platform that aims to improve the RSPO standards by developing and demonstrating criteria that goes beyond the standard RSPO criteria.

Labelling scheme covered is **Singapore Green Label Scheme (SGLS)** which was launched in 1992 and administered by the Singapore Environment Council. Labelling schemes have multiple criteria covering life cycle considerations and require submission of documents proving third-party verification. The SGLS certifies a wide range of products including palm oil and paper products.

We rated the standards based on four aspects:

(1) Criteria: How well the standards' criteria fits the practices for haze-free palm oil production in terms of avoidance and early suppression of fire, protection for forest and peat and minimizing land conflict.

(2) Impact: The impact of the standard in terms of how widely is the standard currently adopted and will potentially be adopted among oil palm growers, including mid-level growers and smallholders. Traceability, while important, was not assessed due to insufficient information on traceability systems in some of the standards.

(3) Trustworthiness: How much consumers can trust that growers really conform to the standard, i.e., whether there are provisions for third-party verification and public scrutiny.

Table 2. Comparison of Regulated Palm Oil Sustainability Standards

	SPO	RSPO NEXT	ISPO	MSPO	POIG	Rainforest Alliance	SGLS
Criteria	3.3 /6	5.4 /6	2.1 /6	2.3 /6	3.9 /6	3.4 /6	3.3 /6
Fires	0.8 /2	1.8 /2	1.0 /2	0.8 /2	0.3 /2	0.8 /2	0.8 /2
Peat	1.0 /2	1.7 /2	0.3 /2	0.3 /2	1.7 /2	0.7 /2	1.0 /2
Forests	0.5 /1	1.0 /1	0.3 /1	0.3 /1	1.0 /1	1.0 /1	0.5 /1
Land conflict	1.0 /1	1.0 /1	0.5 /1	1.0 /1	1.0 /1	1.0 /1	1.0 /1
Impact	3.0 /5	1.0 /5	3.0 /5	2.0 /5	1.5 /5	1.0 /5	1.0 /5
Trustworthiness	4.0 /5	4.5 /5	2.0 /5	1.0 /5	1.5 /5	3.5 /5	0.0 /5
Improvement	4.0 /4	3.0 /4	2.0 /4	2.0 /4	1.0 /4	4.0 /4	1.0 /4
Total	14.3 /20	13.9 /20	9.1 /20	7.3 /20	7.9 /20	11.9 /20	5.3 /20

(4) Improvement: How much the processes and criteria can be expected to improve based on future commitments and actual improvements implemented in the past.

The results of the assessment are shown in Table 2.

Overall assessment

Among the regulated standards, RSPO is the best overall. MSPO and ISPO hold much potential to effect change as national mandatory standards. However MSPO and ISPO need a lot more improvement to be credible standards. SGLS lacks many key elements of a robust certification system, but it is useful as an additional label on top of RSPO certification.

Criteria

RSPO Next has by far the strongest criteria, with POIG, Rainforest Alliance, RSPO/SGLS having average criteria and ISPO and MSPO having weak criteria. We assume that SGLS criteria for sourcing is RSPO certification and so have same scores.

Most standards ban or restrict the use of fire for clearing land, although only RSPO Next requests growers to extend their fire prevention, monitoring and control efforts beyond their plantations and ensure collaboration with local communities and authorities. POIG only prohibits use of fire on peat.

RSPO Next and POIG have strong standards for peat, with a ban on new development on peat on any depth. RSPO/SGLS allows limited development on peat of 100 hectares or 20 percent of the planted area, whichever is higher. ISPO allows development of peat up to three meters deep, while MSPO has no restrictions on new development on peat. Rainforest Alliance does not mention peat explicitly. Peat swamp forests will be protected under the protection of natural ecosystems, but there are no provisions covering non-forested peatland and existing plantations on peat. All other standards mandate that water levels should be managed in existing plantations on peat.

RSPO Next, POIG and Rainforest Alliance have strong protection for forests, with secondary forests protected as well. RSPO/SGLS protects primary forests and high conservation value (HCV) forests, which leaves out secondary forests which are not HCV. ISPO relies on forest zoning and the Moratorium on New Forest Concessions, which have been known to shift boundaries in order to accommodate plantations.[42] For MSPO, protection is based on vaguely defined "high biodiversity value and environmentally sensitive areas".

All criteria except ISPO have strong protections for local communities to minimize land conflict, including use of FPIC. ISPO states that local communities' land rights have to be respected, although there is no explicit mention of FPIC.

Impact

RSPO is the standard with the highest current and potential impact. It is the most widely adopted and recognized certification standard globally. As of August 2017, 19 percent of global crude palm oil is certified by RSPO.[43] RSPO is also exploring Jurisdictional Approach in South

---

[42] awas MIFEE. (2014). Oil palm companies redraw Indonesia's Forest Permit Moratorium Map.

[43] Roundtable on Sustainable Palm Oil (RSPO). (2017). Impacts.

Sumatra, Central Kalimantan, Sabah and Ecuador.[44] RSPO criteria regarding new development also applies to all new plantations for RSPO members, regardless of whether there is intention for these plantations to be certified.

For RSPO however, smallholder inclusion is a challenge, with just 12 percent of RSPO certified area being smallholders although smallholders produce 40 percent of world's palm oil.[45] Challenges faced by smallholders include lack of knowledge of what the RSPO's Principles and Criteria are and how to comply with them.[46] Fortunately, this is an issue that RSPO is actively trying to address. For example, the RSPO Smallholder Support fund provides up to 100 percent of certification cost and member companies provide technical support and monitoring for the smallholders.

As mandatory national standards, ISPO and MSPO have a high potential impact, although current certification rates are still low. ISPO is mandatory for all plantation companies and voluntary for smallholders, but only 12 percent of planted area had been certified as of April 2017.[47] MSPO is mandatory for all oil palm plantations, including smallholders, but only 4 percent of planted area had been certified as of January 2017.[48]

RSPO Next, POIG and Rainforest Alliance have only a small number of plantations or companies certified, and none in Southeast Asia. As of September 2017, there were no products that had obtained the SGLS palm oil label.[49]

RSPO is the most commonly available sustainability certification for palm oil in Singapore. There are at least three suppliers which can provide RSPO-certified cooking oil to eateries and six brands of retail-size RSPO-certified cooking oil (see Table 3).

---

[44] MacIsaac, T. (2017). Jurisdictional certification approach aims to strengthen protections against deforestation. *Mongabay*.

[45] Roundtable on Sustainable Palm Oil (RSPO). (2017). Impacts.; Roundtable on Sustainable Palm Oil (RSPO). (2017). Smallholders.

[46] Shah, V. (2015). Palm oil's big issue: Smallholders. *Eco-Business*.

[47] Ribka, S. (2017). Only 12% of Indonesia's oil palm plantations ISPO certified. *The Jarkarta Post*.

[48] Bioenergy International. (2017). Malaysian palm oil industry gearing up for mandatory MSPO certification.

[49] Singapore Environment Council. (2017). Singapore Green Labelling Scheme (SGLS) Directory.

Table 3. Sustainable Palm Oil Supply in Singapore, September 2017

Company	Can Supply in 15–20 Kg Tins for Eateries	RSPO-Certified Retail Brand(s) (If Any)
Ngo Chew Hong Edible Oil Pte Ltd (Parent Company: Mewah Group)	Yes	Cabbage
Sime Darby Plantation Sdn Bhd	Yes	Hand, Chief, King Rooster
Hap Seng Edible Oils Pte Ltd	Yes	—
Lam Soon Edible Oils Sdn Bhd	No	Golden
Goh Joo Hin Pte Ltd	No	New Moon

Rainforest Alliance and SGLS may however benefit from higher consumer awareness of their respective labels due to the wide range of products bearing their labels. SGLS labels over 3,000 products, which would make it easier for consumers in Singapore to recognise eco-friendly products rather than learn to recognize different labels for different types of products.

In terms of actual impact, there is unfortunately a lack of academic studies measuring the impact of these standards on fire incidence.

RSPO's willingness to suspend errant companies has been effective in compelling these companies to improve their practices. The suspension of Industrial Oxygen Incorporated (IOI)'s RSPO certification in March 2016 triggered an 18 percent drop in IOI's share price and led to suspension of procurement contracts from 27 buyers. IOI thus announced it was focusing on addressing its sustainability issues.[50]

Trustworthiness

The level of trustworthiness varies widely. RSPO/RSPO Next and POIG are multi-stakeholder groupings, with a strong presence of environmental and social NGOs that provide a check and balance. SAN includes non-profit conservation organizations among its members. ISPO, MSPO and SGLS lack independent, third party oversight.

---

[50] Chain Reaction Research. (2017). The Chain: IOI Corporation commits to improving its supply chain risk management.

Transparency is very high for RSPO and RSPO Next, with the following publicly available on website: standard, list of companies certified, audit forms, grievance system, concession maps (with exception of Malaysia but including Sabah), board members. Most standards at least have their criteria publicly available except for MSPO which only provides broad principles and SGLS which does not have any principles or criteria publicly available.

RSPO/ RSPO Next have a transparent grievance system in place that has led to suspensions of errant companies and auditors.[51] In Rainforest Alliance, ISPO and MSPO, grievance system exists but lacks transparency of grievances made. POIG and SGLS do not have grievance system. POIG lacks accreditation, while SGLS relies on peer review.

Challenges remain regarding audit system which may not be able to detect all cases of non-compliance and are subject to conflict of interest because audit companies are hired by the growers.[52] RSPO has tried to address this issue by appointing an accreditation body to check on the auditors.

SGLS does not take into consideration uncertified products by same company. This potentially allows for a two-tiered system whereby the same company produces certified products for more discerning markets while causing environmental and social damage in uncertified areas.

Improvement

RSPO and Rainforest Alliance both commit and have implemented a review and update of their criteria every five years. MSPO also commits to a review every five years, but there is no sign of a review process being implemented even though criteria is up for renewal on 2018. None of the other standards commit to a timeframe for review and update of criteria although ISPO did revise its regulations once in 2015 and POIG updated its indicators once in 2016.

---

[51] Chain Reaction Research. (2017). The Chain: IOI Corporation commits to improving its supply chain risk management.

[52] Environmental Investigation Agency UK Ltd & Grassroots. (2015). Who watches the watchmen?

RSPO and Rainforest Alliance both have systems for review and improvement of procedures and processes. RSPO has numerous working groups focusing on different issues, many of them arising from resolutions adopted at Annual General Meetings. The Complaints and Appeals procedures for example were updated in 2017.

Rainforest Alliance has a learning and support programme to help members and improve certification systems and processes. There is also a SAN Quality System, which tries to improve consistency and transparency in all processes making up the SAN/Rainforest Alliance certification system. The standard contains a continuous improvement system that requires farms to gradually increase their compliance over a six-year period.

No Deforestation, No Peat and No Exploitation (NDPE)

No Deforestation, No Peat and No Exploitation started as a set of principles born out of NGOs' demands for an absolute end to these three damaging practices. In 2009, Nestle was the first company to make an NDPE pledge.[53]

Unlike the standards assessed earlier, NDPE does not have a regulatory body and there is no single standard set of criteria and indicators. However, the broad principles are similar and there is increasing consensus on the best practices and specific criteria to meet those principles.

We assessed the basic NDPE policy with the same methodology applied for regulated standards and the results are shown in Table 4 in comparison with RSPO. Note that for NDPE, the actual criteria, processes and implementation differ from company to company, so the score would differ correspondingly.

Criteria

While there is no one standard set of criteria and indicators, companies committing to NDPE and NGOs pushing for NDPE generally have consensus on a few essentials[54]:

---

[53] Poynton, S. (2014). Wilmar's "no deforestation" goal could revolutionise food production. *The Guardian*.

[54] Aidenvironment. (2017). Nordic investments in banks financing Indonesian palm oil.

Table 4. Comparison of RSPO and NDPE Standards

	RSPO		NDPE	
Criteria	3.3	/6	4.2	/6
Fires	0.8	/2	0.5	/2
Peat	1.0	/2	1.7	/2
Forests	0.5	/1	1.0	/1
Land conflict	1.0	/1	1.0	/1
Impact	3.0	/5	2.5	/5
Trustworthiness	4.0	/5	1.5	/5
Improvement	4.0	/4	4.0	/4
Total	14.3	/20	12.2	/20

- No clearance of High Conservation Value (HCV) and High Carbon Stock (HCS) forests (using the High Carbon Stock Approach).
- No new development on peat of any extent.[55]
- Recognizing the right of local communities to give or withhold their Free, Prior and Informed Consent (FPIC) to any new developments.
- Complying with the fundamental conventions of the International Labour Organization (ILO) and upholding the wider United Nations Guiding Principles on Business and Human Rights.

Many companies also have additional criteria similar to the ones as follows which are relevant to haze:

- No burning for land clearance.
- Best management practices on existing plantations on peat.
- Existing plantings on peat assessed by experts to be unsuitable for replanting will be rehabilitated to original vegetation and conserved.

The exact criteria used and its specificity therefore influences how relevant the NDPE policy would be to preventing haze.

---

[55] Definition of peat differs. RSPO defines tropical peat as "organic soils with 65% or more organic matter and a depth of 50 cm or more" in RSPO manual on best management practices (BMPs) for existing oil palm cultivation on peat (2012).

Combining the two aforementioned sets, we thus derive a basic NDPE policy template that would be relevant to haze:

Criteria

1. No Burning
   a. No land clearance using fire
2. No Deforestation
   a. No development of High Carbon Stock (HCS) forest as determined using the HCS Approach
   b. No development of High Conservation Value (HCV) area
3. No Peat
   a. No new development on peat of any extent, where peat is defined at least as "65% or more organic matter and depth of 50 cm or more"
   b. Best management practices for existing plantations on peat
   c. Existing plantings on peat assessed by experts to be unsuitable for replanting will be rehabilitated to original vegetation and conserved
4. No Exploitation
   a. Respect rights of indigenous and local communities to give or withhold their free, prior and informed consent (FPIC) to any new developments
   b. Complying with the fundamental conventions of the International Labour Organization (ILO) and upholding the wider United Nations Guiding Principles on Business and Human Rights

Scope

- Apply at parent company level, including all of its subsidiaries, for all upstream and downstream palm oil operations that it owns, manages, or invests in, regardless of stake
- Apply to all third-party suppliers that it purchases from or has a trading relationship with

---

(*Continued*)

(Continued)

Processes
• Have grievance mechanism • Make publicly available list of suppliers and mills • Active engagement of suppliers to verify compliance and take corrective action in case of violation

In comparison with RSPO, NDPE has stronger protection for forests and peatland, by banning clearance of all secondary forests, and also all new developments on peat.

Impact

Chain Reaction Research reported in February 2017 that 365 companies globally implemented deforestation-free or NDPE policies.[56] A study in 2016 estimated that 60 percent of global palm oil production was covered by NDPE policies.[57]

For traders and processors, the policy is generally expected to cover not only a company's own plantations, but also third-party suppliers, giving it more reach than traditional certification standards, including RSPO. Preferably (but not always) they should also apply to other commodities and forestry, and the other businesses of company directors.

Also, palm oil supply chain has few traders and processors, of which the largest have adopted NDPE policies. So now lots of growers are confronted by these few traders and processors when these growers may not even be RSPO members.

Whether NDPE policies adopted by traders and processors have a significant impact on the ground depends greatly on the pace of suppliers engagement to ensure their compliance. In mid-2014, Kencana Agri was poised to clear dense forest in Indonesia. Wilmar, which owned a

---

[56] Chain Reaction Research. (2017). The Chain: Indofood Agri Resources new palm oil policy.

[57] Chain Reaction Research. (2016). The Chain: Most global palm oil trade covered by zero-deforestation.

20 percent stake in Kencana Agri, subsequently engaged with them, which resulted in Kencana Agri launching its NDPE policy and sparing the forest.[58]

Actual implementation will take time especially when third-party suppliers, middlemen and smallholders are involved. Challenges include traceability and illegal logging by third parties.[59]

Because the NDPE criteria is not standardized, it is difficult for governments and consumers to base their procurement and buying decisions on them.

Trustworthiness

Companies are expected to have their own grievance system, traceability and transparent reporting. This presents a challenge for smaller companies which lack such capacity. There is also inconsistency in the information made public. For example, Sime Darby reports on hotspots occurring within their concessions, but not on grievances, while Wilmar lists details of all grievances, but does not report on hotspots. As a result, the level of transparency and independent third-party oversight varies considerably.

Third-party oversight and audits are sometimes provided by NGOs or consultants that the company appoints, but at least one company has been accused of misleading these third-party organisations.[60]

When a supplier is found to be non-compliant, the company is expected to ask the supplier to address the issue and terminate the supplier if it fails to do so.

Improvement

Consensus building on the definition and methodology of "no deforestation" has been an area of active work. Initially there were two approaches: HCS+ and HCS Approach, which was converged on 8 November 2016.

---

[58] Aidenvironment. (2016). Impacts of no-deforestation policies.

[59] The Rainforest Alliance. (2013, 2015). An evaluation of Asia Pulp & Paper's progress to meet its forest conservation policy; and additional public statements.

[60] Hance, J. (2016). WWF and Greenpeace break with Indonesia's pulp and paper giant. *The Guardian*.

The HCS Approach steering group is working toward converging HCV, HCS and FPIC.[61]

Access to market as a leverage

Palm oil buyers driving the change

Palm oil sustainability standards create the opportunity to provide economic incentives and disincentives for growers by linking their access to consumer markets with compliance with such standards.

When palm oil buyers such as traders, product manufacturers, retailers and food-service companies commit to sustainable sourcing, it helps push the palm oil growers toward adopting sustainable policies or face the risk of losing their customers. In 2010 Nestle, for example, adopted an NDPE policy, which in turn pushed one of their suppliers, Golden Agri Resources, to also adopt an NDPE policy.[62]

Nevertheless, the take-up rate of certified products and adoption of NDPE policies is still low. Sales of RSPO-certified palm oil is only about half the amount produced.[63] Only 14 out of 55 Consumer Goods Forum (CGF) members assessed by Forest 500 in 2016 had zero or net-zero deforestation commitments across all commodities.[64]

As of September 2017, five eateries in Singapore claim to be using RSPO-certified sustainable palm oil which is a drop in the bucket for this food-loving nation.

While availability of certified sustainable palm oil was previously a challenge in Singapore, this is no longer the case, at least for cooking oil. RSPO is the most commonly available sustainability certification for palm oil in Singapore. As of September 2017, there were at least three suppliers which can provide RSPO-certified cooking oil to eateries and six brands of retail-size RSPO-certified cooking oil.

---

[61] Greenpeace. (2016). HCS convergence process concludes.

[62] Cheam, J. (2011). Golden Agri adopts no deforestation policy. *Eco-Business*.

[63] Roundtable on Sustainable Palm Oil (RSPO). (2017). Impacts.

[64] Bregman, T. P., Mccoy, K., Servent, R. and Macfarquhar, C. (2016). Turning collective commitment into action: Assessing progress by Consumer Goods Forum members towards achieving deforestation-free supply chains.

Civil society raising awareness

Based on our outreach experience, a major barrier is low awareness among the public as well as businesses about the presence of palm oil in the products they use, as well as to option of using sustainable palm oil. Since May 2017, PM Haze has been reaching out to eatery owners/ managers. The majority of eateries PM Haze has reached out to referred to the oil they were using as "vegetable oil". Most of them were also not aware of the link between haze and the unsustainable palm oil they used for cooking. In comparison, when speaking to eateries that did not use palm oil, the majority knew the exact ingredient of the cooking oil they were using. NGO-led engagement with businesses therefore has played an important role in raising awareness. Three of the five eateries that use certified sustainable palm oil in Singapore switched in 2017 after awareness raising by NGOs WWF Singapore and PM Haze.

Efforts to raise awareness of palm oil issues in Singapore have been ongoing, but with slow progress. One of the tools used to generate more public attention and pressure on companies has been scorecards. On 21 September 2017, WWF launched a Palm Oil Scorecard for companies in Singapore and Malaysia, while simultaneously launching a petition against non-responding companies. As a result, six of these companies made a commitment to source sustainable palm oil.[65]

Scorecards can also be useful for consumers and investors to evaluate the companies that they are buying from. They can also serve as a tool allowing companies to compare their actions with those of their peers. However, the scorecards are not comprehensive and only evaluate the larger companies.

Consumer demand, supporting the switch

By avoiding overconsumption of vegetable oil, consumers can reduce pressure on producers to clear land to meet the demand. Consumers can choose to certified sustainable palm oil products to influence businesses toward haze-free practices.

---

[65] WWF. palmoil.sg. (2017). Retrieved https://palmoil.sg/, accessed 18 October 2017.

In Singapore, businesses which have switched to using sustainable palm oil such as IKEA are those which are already marketing themselves as being sustainable. However, many businesses are not willing to use sustainable palm oil even after gaining awareness. Challenges cited by businesses are price and lack of consumer demand.[66]

Based on our exchange with one of the biggest local cooking oil suppliers, the premium for RSPO-certified cooking oil is 10 percent or less, which amounts to about two Singapore dollars per tin, depending on volume purchased.[67] A typical Nasi Lemak stall uses one tin (18 kilograms) of cooking oil for 400 to 600 plates[68] and it costs about two Singapore dollars more per tin to switch to RSPO-certified cooking oil, which comes down to less than one cent per plate.

Consumer demand would therefore need to increase sufficiently to offset the additional cost. Consumers can also urge businesses to be more transparent and adopt sustainable practices. For example, in October 2017, two teenage students ran a successful online petition urging two eatery chains to switch to sustainable palm oil.[69]

Green procurement

While there are no data available on amount of palm oil used by the public sector in Singapore, a study done for UK showed that public sector accounted for 7 to 12 percent of total palm oil and palm kernel oil imports in the UK in 2009.[70]

National governments can thus set an example and shift a significant proportion of the market by enacting green procurement policies which cover palm oil. For example, the UK government's buying standard for food and catering services mandates that "all palm oil (including palm

---

[66] WWF-Singapore & WWF-Malaysia. (2017). Palm oil buyers scorecard — Malaysia and Singapore 2017.

[67] Personal communication.

[68] Personal communication.

[69] SOS Against Haze. (2017). Petition: Tell Old Chang Kee & Polar Puffs to stop frying our rainforests. Retrieved https://www.change.org/p/old-chang-kee-tell-old-chang-kee-polar-puffs-to-stop-frying-our-rainforests.

[70] Proforest. (2011). Mapping and understanding the UK palm oil supply chain.

kernel oil and products derived from palm oil) used for cooking and as an ingredient in food must be sustainably produced."[71] The Singapore government has a public procurement policy that includes printing paper[72] but does not cover cooking oil, even though the public sector procures food via caterers for military camps, prisons, hospitals and ad-hoc events.

National capacity building

National and regional governments and industry associations can also encourage businesses to use sustainable ingredients. Eleven European countries have made national commitments toward 100 percent certified sustainable palm oil.[73] These national commitments are led by industry, government, or both, and actively promote sustainable palm oil among businesses and the public.

In Singapore, the Singapore Alliance for Sustainable Palm Oil (SASPO) aspires to play a similar role,[74] although a nationwide time-bound commitment has yet to be made. In the context of the roadmap for a "Transboundary Haze-Free ASEAN by 2020", Singapore government could work with SASPO to make a time-bound national commitment to 100 percent sustainable palm oil.

Trade-related measures/trade agreements

Finally, national governments can use trade-related measures to influence palm oil producers to improve their standards. Under World Trade Organization rules, members can adopt trade-related measures aimed at protecting the environment, subject to certain specified conditions.[75] The General Agreement on Tariffs and Trade (GATT) has an article on

---

[71] Department for Environment Food and Rural Affairs. (2015). The government buying standard for food and catering services.

[72] National Environment Agency. (2017). Public Sector Taking the Lead in Environmental Sustainability (PSTLES).

[73] Roundtable on Sustainable Palm Oil (RSPO). (2017). National commitments.

[74] Cheam, J. (2016). WWF launches new Singapore alliance on sustainable palm oil. *Eco-Business*.

[75] World Trade Organization. (2017). WTO rules and environmental policies: Introduction.

"General Exceptions" which allow for trade-related measures "necessary to protect human, animal or plant life or health" as long as such measures are not applied in a manner that discriminates between countries where the same conditions prevail, or serves as a disguised restriction on international trade.[76]

The European Union's (EU) stance on illegal timber is a prime example of how trade-related measures have helped to improve standards in the producing country. By banning illegal timber from entering the EU while simultaneously working with timber producing countries to implement systems for verifying timber legality, the EU helped to motivate Indonesia to strengthen the regulations and implementation of its national mandatory timber legality certificate.[77]

As chairman of ASEAN in 2018, Singapore will have the opportunity to coordinate with other ASEAN member states to develop a common palm oil certification system for ASEAN.

Access to finance as a leverage

Responsible financing

Financial institutions and retail investors can influence companies in forest-sector, by tying access to loans or investments with compliance with sustainability standards. This is commonly known as socially responsible investment (SRI) or simply "responsible financing".

Responsible financing would require both negative screening and positive screening. Negative screening involves eliminating companies which adopt objectionable, unsustainable practices.

An example of a financial institution which actively monitors and addresses sustainability risks in its investments is Norges Bank Investment Management (NBIM), which manages Norway's 900 billion US dollars sovereign wealth fund, the Government Pension Fund Global (GPFG). This is the world's largest sovereign fund. Between 2012 and 2015, NBIM

---

[76] General Agreement on Tariffs and Trade. (2012). Article XX: General Exceptions.
[77] EU FLEGT Facility. (2017). Indonesia-EU Voluntary Partnership Agreement. Retrieved http://www.euflegt.efi.int/publications/indonesia-eu-voluntary-partnership-agreement, accessed 26 September 2017.

divested more than 30 palm oil companies due to high risk of contributing to tropical deforestation.[78]

In Singapore, the Association Banks in Singapore (ABS) issued Guidelines on Responsible Financing in October 2015, with the expectation that banks are expected to fully comply with the guidelines by 2017. The guidelines define the minimum standards on responsible financing practices to be integrated into banks' business models in Singapore. DBS, OCBC and UOB had announced in their 2016 Annual Reports their responsible financing frameworks, which embed environmental, social and governance factors (ESG) into deciding who they lend money to and what conditions are included in the loan.

However, only DBS has set standards for palm oil sector. DBS has specified that new borrowers should "additionally demonstrate alignment with no deforestation, no peat and no exploitation policies ... We will also consider new customers who have achieved RSPO certification or are able to demonstrate that they are working towards achieving RSPO certification within a satisfactory timeframe."[79]

Other financial institutions in Singapore such as our Singapore-owned institutional investors Temasek Holdings and GIC, insurers and university endowment funds have yet to announce policies for responsible financing. Some banks such as Standard Chartered[80] and HSBC[81] have palm oil policies based on both NDPE and RSPO with clearly defined standards and timeframes with expectations for both new and existing clients.

Financial institutions have the scale to implement the NDPE policy and should already be doing risk screening for companies they loan or invest in. RSPO certification can be used to supplement as a way to improve verification and tap on their grievance procedure. HSBC for

---

[78] Regnskogfondet. (2017). World's largest sovereign wealth fund: Palm oil sector still too risky for investment.

[79] DBS Bank. (2017). Our approach to palm oil sector. Retrieved https://www.dbs.com/sustainability/responsible-banking/responsible-financing/our-approach-to-palm-oil-sector/default.page, accessed 26 September 2017.

[80] Standard Chartered. (2017). Standard Chartered position statement: Palm oil. Retrieved https://www.sc.com/en/resources/global-en/pdf/sustainabilty/Palm_Oil_Position_Statement.pdf, accessed 19 October 2017.

[81] HSBC. (2017). Statement — Revised agricultural commodities policy: Palm oil.

example informed RSPO about allegations that one of its customers was gearing to clear pristine rainforest, thus triggering an investigation by RSPO.[82]

Financial institutions that need capacity-building support can turn to groupings such as United Nations Environment Programme — Finance Initiative, Banking Environment Initiative (BEI), UN Principles for Responsible Investment (PRI) and Soft Commodities Compact. NGOs can also play an important role in capacity building for businesses. For example, the Singapore Institute of International Affairs (SIIA) created the Collaborative Initiative for Green Finance in Singapore to explore green financing possibilities for banks and come up with a framework to assess green practices of borrowers.[83]

Another way finance can help is by supporting sustainable activities. This is commonly known as positive screening. National and international bodies can also provide incentives and frameworks to support growth of responsible and green finance that incorporates positive screening. Under the Sustainable Shipment Letter of Credit by the BEI, the International Finance Corporation will offer preferential terms of credit to its partner banks when they finance the import of RSPO-certified palm oil to emerging markets.[84]

Indonesia has a "Roadmap for Sustainable Finance in Indonesia" to determine which measures need to be taken to improve the sustainability of finance in Indonesia, and to have these implemented by 2024.[85] Singapore and ASEAN as a region have yet to develop such initiatives.

Sustainability reporting

The key to successful use of finance as a leverage is transparency — for both palm oil companies and the financial institutions. Increased

---

[82] Paddison, L. (2017). HSBC triggers investigation into palm oil company over deforestation allegations. *The Guardian.*

[83] Othman, L. (2017). Framework being developed for banks to assess green practices of borrowers. *Channel NewsAsia.*

[84] Pek, S. (2016). How can banks spur the palm oil industry toward sustainability? *Mongabay.*

[85] Otoritas Jasa Keuangan. (2014). Roadmap for Sustainable Finance in Indonesia.

transparency of financing policies and clients helps stakeholders of the financial institution have oversight on how well responsible financing is being implemented, and sends a market signal to potential creditors that access to finance is increasingly tied to their sustainability performance.

NGOs have played a role in increasing transparency by conducting research and publishing research on the financing practices of financial institutions. Sustainability reports give information about the company's environmental, social and governance performance. However in 2014 and 2015, only 37.1 percent of 502 companies on the Singapore Stock Exchange (SGX) reported on their sustainability performance.[86]

National stock exchanges have mandated sustainability reporting for listed companies, e.g., Indonesia's IDX, Malaysia's BURSA and Thailand's SET. In Singapore, sustainability reporting will be on a "comply or explain" basis from financial year ending on, or after 31 December 2017.[87] However, the SGX guidelines currently do not require independent assurance to verify the claims on the sustainability report and do not specify factors which a company should report on.

Summary of recommendations

Intergovernmental collaboration

(1) Strengthen environmental safeguards in trade agreements/deals.
(2) Singapore as chairman of ASEAN in 2018, to promote responsible finance and green micro-credit e.g., financing adoption of sustainable palm oil.
(3) Singapore as chairman of ASEAN in 2018 can coordinate with other ASEAN member states to develop a common palm oil certification system for ASEAN.

---

[86] Loh, L., Nguyen, T. P. T., Sim, I., Thomas, T. and Wang, Y. (2016). Sustainability reporting in Singapore: The state of practice among Singapore Exchange (SGX) mainboard listed companies 2015.

[87] Loh, L., Nguyen, T. P. T., Sim, I., Thomas, T. and Wang, Y. (2016). Sustainabilty reporting in ASEAN: State of progress in Indonesia, Malaysia, Singapore and Thailand 2015.

Singapore government

(1) Government green procurement policy should mandate that all palm oil used in cooking oil must be sustainably produced, with RSPO-certification as the current acceptable standard.

(2) Government support for sustainable palm oil capacity-building by:

  (a) Making a time-bound national commitment to use sustainable palm oil;

  (b) Supporting SASPO and provide grants to support the transition.

(3) Singapore government to set up the preconditions, such as proper strategies, technologies, incentives and regulations to promote responsible finance incorporating both negative and positive screening, e.g. eliminating customers causing deforestation and financing adoption of sustainable palm oil, as well as forest- and peat-friendly agriculture.

(4) SGX to put in a timeframe for listed companies to improve their sustainability reporting standard, e.g., having third-party assurance, stakeholder consultation.

Civil society groups and general public

(1) NGOs to scale up awareness outreach among public and businesses about sustainable palm oil and responsible finance.

(2) NGOs to promote transparency and public scrutiny by analysing:

  (a) Financing policies of major financial institutions and funds in Singapore;

  (b) Palm oil procurement policies of major eateries, manufacturers and retailers;

  (c) NDPE policies and their implementation among palm oil companies which are listed and/or have headquarter in Singapore.

(3) Academics and NGOs to regularly review and provide recommendations on palm oil standards as they evolve.

(4) NGOs and consumers to urge businesses which use palm oil to switch to RSPO-certified palm oil, with a focus on urging businesses

with a presence beyond Singapore to use RSPO-certified palm oil across all global operations.

(5) Consumers to support RSPO-certified palm oil products; eateries that use no oil or RSPO-certified palm oil; and retailers that use RSPO-certified palm oil for their housebrand palm oil products.

(6) Consumers, businesses and other organizations should reduce consumption of fried food.

(7) Shareholders to urge listed palm oil companies, buyers and financial institutions to use RSPO-certified palm oil and/or adopt robust NDPE policy.

Eateries, manufacturers and retailers that use palm oil

(1) Adopt a time-bound plan for 100 percent of palm oil used for cooking oil across the company's global operations to be RSPO-certified.

(2) Communicate and educate their customers about the use of sustainable palm oil products.

Palm oil growers, traders and processors

(1) Adopt and implement a robust NDPE policy.

(2) Have a time-bound commitment for 100 percent RSPO certification for own plantations.

(3) Report annually on their progress on NDPE and RSPO certification in their sustainability/annual report.

Financial institutions

(1) Main financial institutions in Singapore* must all adopt ESG policies with publicly disclosed sector-specific policies covering Agriculture and Forestry that requires customers who are palm oil growers, traders and processors to:

  (a) Adhere to NDPE;

  (b) Have RSPO membership and a time-bound plan for 100 percent RSPO certification for own plantations.

(2) Main financial institutions should publish a list of clients they lend money to in high risk sectors including agriculture and forestry.

*Scope: Main financial institutions

- Local banks: OCBC, DBS, UOB.
- Singapore-owned institutional investors: GIC and Temasek.

15 Partnerships and Long-Term Solutions Will Be Critical

Jose Raymond

Former Chief Executive, Singapore Environment Council
Vice-President (Corporate Affairs), Asia Pulp and Paper
jose.raymond@spinworldwide.org

Introduction

In June 2013, when Singapore was enveloped in a cloud of haze, I contributed an opinion piece to Singapore daily, TODAY. At the time of the contribution, I was the Chief Executive Officer of the Singapore Environment Council, a non-profit organization which champions environmental causes in Singapore.

The article, titled "Long-Term Fix to Haze Must Involve 3Ps", which I shared my views on why a multi-pronged partnership approach across as many sectors as possible would be critical in ensuring that the almost annual issue of haze would not hound Singaporeans for a lifetime.

Four years after I contributed the opinion piece to TODAY, and particularly so after spending almost a year in Asia Pulp and Paper as its Vice-President of Corporate Affairs, one of the world's largest paper manufacturers who have been accused of the haze, I feel a lot stronger about the need for greater collaboration across as many sectors, and governments, if we are to think about finding a long-term solution to the haze issue which has plagued the region almost annually for the last three to four decades.

Till today, I am often asked why I made the decision to join APP, especially after the company received a notice from the National Environment Agency (NEA) under Sections 9 and 10 of the Transboundary Haze Pollution Act (THPA) for the haze episode in 2015. The THPA came into effect on 25 September 2014, which made it possible for the Singapore government to take action on any company based in Singapore or overseas which causes pollution in Singapore.

My response has been that it is always better to help an organisation like APP which has already started on its journey towards sustainability, and support the management of the company be the change from within especially since it has already made a concerted effort to change the way it conducts its business. The APP announced its Forest Conservation Policy in early 2013, and since then, it has made steady progress towards heading on the road to sustainability. Even Greenpeace acknowledged the progress made by APP, when it announced in October 2013 that it was suspending active campaigning against the company following the release of the Forest Conservation Policy.

What causes the haze?

There are multiple reasons why fires are started in agricultural areas in Indonesia, and also in Malaysia and in parts of Indochina.

Extensive land clearance and cultivation, illegal logging and the burning of forests to clear land for cultivation are primarily the main causes of fires in Indonesia and in the region. Forest fires often destroy high capacity carbon sinks, including old-growth rainforest and peatlands.

While there has been extensive outreach to poor, rural communities in Indonesia on the harm of clearing through slash and burn, burning will remain a preferred land-clearing method as it is quick and efficient, requires minimal labor while enriching the soil.

Apart from land clearance for agriculture, communities are also entitled to land spaces within concessions granted for their own livelihood. For these communities, the only way they know how to clear land would be through burning. How many such communities can one find within the concessions? It varies and for example, APP's suppliers have reported as having between 20 to 30 communities living within concession areas at any given time. Sometimes, new villages just spring up overnight, which makes policing even harder on the ground.

Apart from communities, there are also accidental fires which are started due to smoking, cooking within the concession areas and also through arson. Arson can be caused by competitors who try to reduce the yield of other companies, or also because of financial compensation due to land claims.

The haze of 2015

From around August 2015 onwards, parts of Sumatra and Kalimantan were ravaged by forest fires for weeks because of the extreme dry season, which was further exacerbated by the El Niño phenomenon.

During that period, as Senior Director of Corporate Communications and Stakeholder Engagement at the Singapore Sports Hub, I was deeply involved during the hazy days when the Pollutant Standards Index (PSI) hit unhealthy levels, as we had to communicate to the general public why we needed to shut a few of the outdoor venues due to the poor air quality. The Singapore Swimming Association, where I hold a position of Vice-President (Finance), also had to cancel a day of competition for the FINA World Cup swimming competition at the OCBC Aquatic Centre. This inevitably led to losses in revenues for the association. Many foreign athletes who had come by to Singapore for the event had to return without having competed at all.

As a result of the prolonged haze, the NEA issued notices to six companies under the THPA to extinguish the fires and to prevent the spread of any fire in land concessions which were occupied by them. Two companies, including APP, responded to the requests for information by the NEA.

But after receiving the information from the companies, the NEA has not been able to continue with action till today because Indonesia's Minister for Environment and Forestry Siti Nurbaya Bakar said that she would have to discuss the matter with members of her Cabinet first.

The need for greater cooperation

What are the chances of Indonesia sharing the requested information with Singapore which would lead to its nationals and companies being prosecuted in Singapore courts? My own discussions with government officials from the Environment Ministries in the region indicate that Association of Southeast Asian Nations (ASEAN) is still many years away from being open enough to share such information with neighboring countries.

Indonesia aside, senior government officials from Thailand and Malaysia whom I have met as part of my current role indicate that their countries are not considering any THPA-like legislation for the time being as they still believe that issues should be solved between government to government.

At the ASEAN level, the ASEAN Agreement on Transboundary Haze, which has been in place since 2002, needs to be an avenue where greater cooperation is established between member nations. The platform is a right one for governments of the region to forge greater collaboration among each other so that there is greater understanding and cohesion in trying to eradicate the issue on a long-term basis.

It will be impossible for Singapore to go at it alone and will surely need the support of the other governments of the region if it wants to further push the envelope on the matter. This point was also made in a report by the Singapore Institute of International Affairs in its special report on the haze entitled "From the Haze to the Resources: Mapping a Path to Sustainability".

While governments steer towards greater collaboration, what is also clearly needed is for all major players and even smallholders who own land concessions to agree to a set of terms which all would abide by.

Based on the way land is governed in Indonesia and the multiple challenges on the ground, it will be useless if only a handful of companies like APP and Golden Agri Resources have good agricultural practices and robust fire prevention strategies on the ground. This is because with neighboring concessions placed almost side by side in some areas, the risk of fires encroaching into each other's concessions will always be prevalent. All companies and concessions owners, regardless large, mid-sized or even small, will need to be bound by the same set of rules of engagement.

Finally, it is also important for non-government organizations (NGOs) to work together with government agencies and the concession owners so that solutions can be found to many of the complex issues which occur at ground zero. While it is important to continue highlighting any errant practices, NGOs and consumer groups will also need to assess if boycotts actually serve the intended purposes of bringing about change.

For many, boycotts will only force more companies into hiding and will stop them from being more open and transparent as they should be. And this would only be counter-productive to the intended objective of having greater transparency in the agro-forestry sector. More importantly, regional NGOs should work together to conduct research campaigns on products and companies so that no one works in isolation or in silos. The issue requires many hands and brains on deck and it will be wiser if NGOs collaborate as opposed to going at it alone.

A Development or Education Fund could be set up and administered by an NGO in Singapore with an Institution of Public Character status. These funds could be put aside by corporations involved in paper and palm oil products, and companies which believe in corporate sustainability and responsibility. These funds could be used to educate not only people in Singapore about what they can do to force businesses to change irresponsible practices, but also help hapless villagers in Indonesia living in areas which are blanketed in smog about this time every year. The NGOs could also tap the funds to engage companies in changing their practices over time.

Conclusion

Striving for a sustainable solution to the haze requires the action and collaboration of the public, private and people sectors, working closely together, now more than ever. And the sooner the various players come together to seek solutions for the long-term good of the region, the faster we will be able to rid the region of the scourge of poor air quality.

References

Channel NewsAsia (2016, March 2), Singapore stresses need for greater accountability on haze to Indonesia, CAN. Retrieved from http://www.channelnewsasia.com/news/singapore/singapore-stresses-need-for-greater-accountability-on-haze-to-in-8112116

Greenpeace (2013). APP's Forest Conservation Policy, Progress Review October 2013. Greenpeace International, Amsterdam, The Netherlands. Retrieved from http://www.greenpeace.org/international/Global/international/publications/forests/2013/Indonesia/APP-Forest-Conservation-Policy.pdf

Raymond, J. (2013, June 25). Long-term fix to haze must involve 3Ps. Today. Retrieved from http://www.todayonline.com/commentary/long-term-fix-haze-must-involve-3ps

16 The Indonesian Transboundary Haze Game: Countering Free-Riding and Local Capture

Ridwan D. Rusli[1]

Professor of International Finance and Strategy,
Technische Hoschschule Köln (University of Applied Sciences Cologne)
ridwan.d.rusli@gmail.com

Introduction

Although costly in terms of economic damages, environmental and opportunity losses, and illegal apart from exemptions given to individual and smaller groups of farmers, man-made forest burnings continue to be pervasive across the larger islands in Indonesia, especially Sumatra and Kalimantan (CIFOR).[2] The so-called slash-and-burn practices are often more profitable than mechanical means of clearing lands for oil palm and

---

[1] The author was lecturer in public economics and industrial organization of NTU. He is grateful to Euston Quah and Yohanes Eko Riyanto for valuable comments and World Bank Jakarta team for clarifications on estimated costs of fire and haze. All errors are the author's.

[2] See multiple studies and publications by Center for International Forestry Research (CIFOR). The Republic of Indonesia's Law No. 32 (2009) on the Protection and Management of Environment and Government Regulation No. 4 (2001) on Management of Environmental Degradation and/or Pollution prohibit the use of burning for land clearing. The government does, however, allow slash-and-burn practices for financially constrained individual and small groups of farmers clearing less than 20 hectares in aggregate.

timber plantation companies, and cheaper for individual and small groups of farmers. This cost advantage of burning is particularly pronounced in heavy forests and peat swamps (Gouyon and Simorangkir, 2002). The resulting transboundary haze pollution, exacerbated by extreme weather events such as El Niño and droughts, has become a health and economic problem for Indonesia and its neighbouring Association of Southeast Asian Nations (ASEAN) countries since the late 1990s. This time period saw oil palm identified by the government of Indonesia as one of the strategic sectors to help promote growth in exports. While forest fires result in direct damages to equipment and infrastructure and direct costs of health treatment and firefighting, the indirect or secondary negative impact of fires and the resulting transboundary haze pollution is much higher. The latter include the costs of subsequent land reclamation and foregone pro-duction revenues (during the replanting period), environmental damages and biodiversity losses, as well as opportunity losses to local, regional and cross-border trade, industry and tourism activities (World Bank, 2016).

Despite repeated pressures from and multiple negotiations and treaties with neighbouring ASEAN governments, international and multilateral institutions, limited progress has been achieved in the abatement attempts to prevent and mitigate forest fires and the resulting haze pollution, as well as in stricter enforcement of zero-burning policies for medium- and large-sized plantation companies and farmer groups (Nguitragool, 2011; Purnomo and Shantiko, 2015; Tan, 2015; World Bank, 2016). After two decades of regional and international efforts, approximately 100,000 fires across 2.6 million hectares of Indonesian lands burned between June and October 2015 (World Bank, 2016). Mandatory certification schemes such as the Indonesian Sustainable Palm Oil initiative, the Roundtable on Sustainable Palm Oil and the Indonesian Palm Oil Pledge face challenges due to the attempts by smaller producers to win exemptions and because of lack of transparent, agreed-upon maps of sensitive areas. Nevertheless, efforts continue for example through the government's peatlands morato-rium and restoration plan and the OneMap policy. Regionally within ASEAN, the Indonesian government has struggled for 12 years to com-plete the ratification of the 2002 ASEAN haze treaty.

Nguitragool (2011) argues that, apart from normative constraints and organizational customs within the ASEAN institution, a major impediment to a more rapid ratification has been domestic politics in Indonesia. In particular, Indonesian government negotiators that are primarily led by the Ministry of Environment lacked the authority, faced ongoing decentralization forces and resistance from the Indonesian legislature. While the Ministry barely wields decision making authority over land and forest policy at the central bureaucratic level, it does not possess the power to enforce domestic laws on land and forest fires. Despite the revised 2009 Law, it lacks the resources and is dependent on provincial governors and local district heads to monitor and ensure compliance at local levels. This lack of regional and local authority is exacerbated by local bureaucrats' resistance to national, let alone foreign, interference in firefighting and more generally in the implementation of abatement and enforcement programs.

Purnomo and Shantiko's (2015) study helps rationalize some of the domestic and local difficulties that impede the central Indonesian government's implementation of appropriate fire and haze abatement and zero-burning enforcement programs. The benefits and profits from slash and burn for land clearing are distributed amongst multiple potential beneficiaries ranging from local business elites, plantation companies, local bureaucrats including provincial governors, district heads, local forestry officials and the local legislature, to farmer and individual landowners engaged in slashing, cutting, and burning activities. Disentangling the cost-and-benefit calculation, and in turn the incentives, of each key player in the local political and business value-chain is difficult. Connecting the political communication and power networks from local bureaucrats to their national counterparts, which are significantly but not always driven by party alliances, adds another dimension to the complex domestic political and business dynamics.

Neighboring governments such as Malaysia and Singapore who have tried to engage and proactively help the central and local governments in Indonesia in their abatement and enforcement programs, have been thwarted by nationalistic and, presumably, indirectly, by local business interests. While continuing to engage the Indonesian government, some of them have identified possible policy leverage against some of the larger and internationally operating plantation companies. Singapore, in particular, has ratified its 2014 Transboundary Pollution Act (Tan, 2015). This

Act is not expected to be able to force the Indonesian government to comply and fulfill its responsibilities under international laws, which range from due diligence to information sharing responsibilities. However, the Act does give Singaporean courts the power to prosecute Singapore-domiciled or Singapore Stock Exchange-listed plantation companies. Nevertheless, difficulties in matching data on burning versus concession areas and in verifying land rights and control over burning practices taking place in multiple areas across Indonesia may still pose challenges to the Singaporean courts.

We utilize as basic set-up a simplified Indonesian haze game to identify the challenges that ASEAN neighboring governments face in negotiating with and helping the Indonesian government devise and implement an all-encompassing fire and haze abatement and zero-burning enforcement programs. Every year, before the dry season starts and particularly in more pronounced El Niño years, the question arises again for the Indonesian and the neighboring governments to discuss and negotiate whether to invest in abatement enforcement, and who should bear the cost to fund such programs. We observe that a central Indonesian government that takes a holistic perspective and considers in its welfare objective all direct and indirect costs of fire and haze finds itself in a repeated game-of-chicken, or a sequential chain-store game, with its neighbors. The classic two-player stage-game of the game-of-chicken is known to have three Nash equilibria: either one player invests in abatement and enforcement unilaterally, or the other, or both play probabilistic mixed-strategies between investing or not investing in abatement and enforcement. However, in practice, the Indonesian central government's efforts are impeded by a strong local collusion between local bureaucrats, business elites and plantation companies who weigh the profitable slash-and-burn practices against local, direct costs of fire and haze. These local direct costs are significantly lower than the indirect and longer-term costs to industry and environment, particularly nation-wide.

Local bureaucrats, business elites and companies have the power and sufficient rents to incentivize farmers and residents to engage in illegal slash-and-burn land clearing practices, while influencing the national authorities to either look away or support their local rent-seeking activities. In this context, local farmers and residents participate in a kind of multi-principle common-agency game where the local bureaucrats, business elites, plantation companies, and the central government, influence them to earn income from illegal burning practices whenever these are more profitable than legal mechanical land clearing. As a consequence, the game-of-chicken shifts to a game with only one pure Nash equilibrium where the Indonesian government, dominated by its local interests, ends up free-riding on the neighboring governments' efforts and investments in abatement and enforcement programs. On the one hand, this is the only domestic consensus the central Indonesian government can gain effective authority to negotiate for and the local support to implement. On the other hand, the reason for the neighboring government's decision to invest in abatement and enforcement is the significant costs and losses to their citizens' health and tourism and industrial sectors as a result of the transboundary haze.

Following an analysis of these free-riding, common agency and capture problems we identify possible solution strategies that the Indonesian and neighboring governments can pursue. Firstly, given the sequential game-of-chicken nature of the negotiations, the one party that initially credibly insists not to invest and fund abatement and enforcement will force the other government to invest and bear the cost of such programs. Secondly, considering that the neighboring government may not be able to effectively, unilaterally invest in and implement the complete abatement and enforcement programs on its own, resulting in residual health and economic costs to everyone, it may be able to force the Indonesian government to jointly invest in abatement and enforcement. This is particularly plausible when the neighbouring government credibly signals its stance of not investing, and together with international and multi-lateral institutions jointly force the Indonesian government to act, or shame the latter into acting. This is akin to solution policies that could shift the payoff structure of the game into one where investing in abatement and enforcement becomes the dominant strategy for the Indonesian government. Here the neighboring government can decide whether to participate, assist or co-invest together with the Indonesian government. Thirdly, strategies must be found to break the collusion between local bureaucrats, business elites and companies, such as to ensure sufficient decision making authority and the power to implement effective abatement and enforcement policies for the Indonesian government. These include offering incentives and transfers for farmers to seek alternative employment opportunities, as well as to small- and medium-sized plantation companies, while strictly monitoring and severely punishing local bureaucrats, business elites and companies for any active involvement, or passive knowledge, of illegal burning activities.

This chapter complements the existing theoretical literature on transboundary pollution control and coalition building. Carraro and Sinisalco (1995) study transfers and simultaneously-linked negotiations that can be used to incentivize collusion-building (in our context, to engage in legal mechanical land clearing activities). Chander and Tulkens (2006) discuss that the need for international cooperation derives from the public goods nature of the externality, and that the source of cooperation lies in cooperative game theory. They then analyze the conditions for stability to counter free-riding and the difficulties in ensuring self-enforcement. Fünfgelt and Schulze (2011), on the other hand, observe that when a government is focused on production and employment, but less on environmental costs, it may be swayed by producers' lobbying and set too low taxes. At the same time, lobbying and shaming by international institutions may result in lower equilibrium pollution levels. Bertinelli *et al.* (2015) use a dynamic games framework with two players where only one player commits to specific abatement policy from the outset. They derive that pollution levels can be lower because a stringent environmental quality target will induce the committed player to produce an abatement effort that more than compensates for the free-riding behavior of the non-committed player. By contrast to the above studies, we use empirical estimates of costs and benefits of Indonesian forest slash-and-burn practices in a simple haze game and infer a repeated game-of-chicken, combined with a multi-principal common-agency setup, and use a simple static-comparable analysis as a basis for possible solution strategies and policies.

Following this introduction and literature overview, we describe game theoretical and common-agency framework to analyze the Indonesian fire and haze problem in the second section. The third section discusses our proposed abatement and enforcement strategies and policies. The fourth section summarizes and identifies areas for future research and policy evaluation.

Game-theoretical framework

Indonesian haze game-of-chicken

Distilled down to its basic features, the transboundary gaming and free-riding problem can be analyzed by way of a two-by-two game. The game comprises two players, the Indonesian and the neighboring government. Each of the two players chooses between two policies, whether to invest in and implement abatement and enforcement policies to target zero-burning, or not. For the not-invest policy, we examine three broad scenarios. In an 'ideal' scenario, the utilitarian Indonesian government calculates aggregate welfare as the sum of the tax revenues it receives from farmers, residents and companies, their incomes, plus the profits of oil palm and timber plantation companies, netted off against the direct and indirect costs of forest fires and haze pollution.[3] Here the Indonesian central government calculates direct costs to include damages to infrastructure and equipment from the forest fires and the cost of health treatment and fire-fighting.[4] Indirect costs include losses from post-fire land reclamation, opportunity loss of production revenues from fires and the negative effects of haze pollution on tourism, transport and industry, the cost of environmental degradation and biodiversity loss, and the secondary effects of residents' fire and haze-related illnesses on their productivity and employment opportunities.

In a more 'realistic' scenario, we assume that the Indonesian central government considers in its welfare objective the tax revenues, incomes and company profits, while subtracting only the direct costs of fires such as equipment and infrastructure damages and the costs of health treatment and fire-fighting, as well as indirect opportunity losses of the regional economy and industry. What it may not consider is the less easily quantifiable cost of environmental degradation and biodiversity loss. In both scenarios we ignore the leakage of economic rents to the local bureaucrats and local business elites, because these transactions are generally non-transparent and

---

[3] Here the Indonesian government acts as a benevolent planner with (quasi) complete information.

[4] See Adriani (2016) and World Bank (2016) for an estimation of costs, damages and losses from the Indonesian forest fires and haze pollution.

unobservable to the central government, and may partly flow back into the local and national economy over time.[5]

Furthermore, a third 'capture' scenario is equally conceivable. Whenever local bureaucrats are captured by business interests and gain sufficient power to impede the central government's abatement and zero-burning enforcement programs and influence the latter's negotiation with neighboring governments, local bureaucrats could force the central government to care only about the direct costs of forest fires and haze. Indeed, local bureaucrats may indeed see the benefits to the local economy of slash-and-burn practices to over-compensate for the direct costs of fire and haze.

By contrast, when the Indonesian government chooses to introduce fire and haze abatement and enforcement policies, its aggregate net payoff will include the relevant tax revenues, incomes and company profits, which will differ from the not-invest policy because plantation and forestry company profits will be different under strict enforcement of zero-burning regulations.[6] From these revenues the government subtracts the costs of prevention and abatement, any fiscal incentives and subsidies offered to farmers for transitioning to other employment alternatives and to companies to incentivize them to shift to zero-burning practices, as well as any costs of enforcing zero-burning regulations and sanctioning local bureaucrats and business elites as well as plantation companies conducting illegal activities.

The group of neighboring governments include Singapore, Malaysia, Thailand and Brunei. They differ in terms of their geographical, economic and political interests. For the purpose of our simple haze game, we use as representative neighboring country government the examples of Singapore or Malaysia, which are closer to the areas affected by the forest fires in Sumatra and Kalimantan, suffer more severe health hazards and economic losses to their population, while holding certain economic interests in the relevant plantation companies. Malaysia is affected by transboundary haze

---

[5]We designate local bureaucrats to include district-level bureaucrats and legislators. Following the 1999 decentralization laws, local districts have gained autonomy, and district heads have been selected through public, direct elections (Duek *et al.*, 2010).

[6]Gouyon and Simorangkir (2002) and Simorangkir (2007) compare the costs of slash-and-burn and zero-burning practices.

and pollution from Sumatra and Kalimantan but harbours a number of major plantation companies, some of which may still be involved in or benefit from the cheaper slash-and-burn practices in Indonesia. Singapore, amongst the most vulnerable to the haze and pollution health problems, houses financial institutions, capital market investors and transport and other supplier industries that serve the regional plantation companies. Brunei and Thailand mainly suffer the health consequences of the haze and pollution but both have limited ability to influence policy in Indonesia, or have lesser economic interest in the regional plantation companies.

In line with past and current observations, the representative neighboring government is assumed to continue offering technical assistance to the Indonesian government, regardless of the latter's decision to invest in and incentivize fire and haze abatement and enforcement, or not. Beyond such lower cost involvement, however, the neighboring government must decide whether to invest in unilateral abatement and enforcement policies, in coordination with but (in the worst case) unilaterally without financial support from the Indonesian government, or alternatively, to jointly invest in and implement zero-burning policies with the Indonesian government. It is intuitive, given the intricate political economy of local bureaucrats, business elites and company interaction as well as natural sovereignty issues, that direct unilateral involvement and investment by the neighboring country are likely to be less effective and often not welcome. In any case, the neighboring government weighs its payoff under unilateral intervention against a joint investment in abatement and enforcement policies together with the Indonesian government. In the unilateral abatement and enforcement case, its aggregate payoff is dominated by the full cost burden of fire and haze abatement and enforcement as well as the costs of any residual health hazards that may be unavoidable given the likely less effective unilateral action without full support of the Indonesian and local governments. In the latter case of a joint investment and full collaboration between the Indonesian center and local governments and the neighboring country government, the costs of abatement and enforcement can be shared and the efficiency of such policies can be significantly improved. By contrast, if both the neighboring government and the Indonesian decide not to invest in abatement and enforcement, the neighboring economy will suffer from direct health costs and indirect economic opportunity losses.

Indonesian government	Neighbouring government	
	Not invest	**Invest in abatement and enforcement**
Not-invest	$[B - H^l, -H^n]$ Indonesia: Taxes & profits (with burning), (-) direct & indirect costs of fires & haze Neighbour: (-) direct & indirect costs of haze	$[Z, -C]$ Indonesia: Taxes & profits (zero-burning) Neighbour: (-) full cost of abatement & enforcement
Invest in abatement and enforcement	$[Z - C, 0]$ Indonesia: Taxes & profits (zero-burning) (-) full cost of abatement & enforcement Neighbour: rely on Indonesian government	$[Z - (1-\gamma)C, -\gamma C]$ Indonesia: Taxes & profits (zero-burning) (-) part of cost of abatement & enforcement Neighbour: (-) part of cost of abatement & enforcement

Figure 1. Indonesia Haze Game

Examining Figure 1, we define B as the benefit of tax revenues from companies and farmers, profits of plantation companies and incomes of residents, farmers and local businesses, all of which the government captures in its welfare objective under the assumption of continued slash-and-burn practices. At the same time Z is to represent the corresponding taxes, profits and incomes under strict enforcement of zero-burning policies. It is reasonable to expect, given the significant cost savings of slash-and-burn practices over purely mechanical land clearing methods with zero-burning in heavy forests and peatlands, that the taxes and profits are higher when at least part of the landbanks are cleared with the help of fires, thus

$$B > Z. \tag{1}$$

This can be inferred from Figure 2, where the estimated profit advantage of land clearing through slash-and-burn practices, $B - Z$, is in the order of magnitude of more than 1.7 billion US dollars.[7] Note that burned lands also require lower crop protection expenditures before planting, while increases in fertilizer costs are likely neglected because of

---

[7]See Figure 2 for rough estimates of the damages and losses from forest fires and haze; the benefits of slash-and-burn practices to company profits (implicitly including the government's tax revenues from profits); residents', farmers' and local business elites' incomes; and a crude estimation of anticipated costs of forest fire and haze abatement and enforcement of zero-burning regulations.

Payoff estimates for the Haze Game (USD billion)				
Cost of forest fire and haze 1)			**Profit benefit of slash-and-burn vs. zero-burning 3)**	
Direct costs and damages	= H°	2,2	Peat swamps	37%	1,2
Indirect: agriculture & forestry losses		6,9	Forests	63%	0,5
Indirect: tourism & industry losses		2,7			
Indirect: environ. & biodiversity losses		4,3	**Higher company profits**	= B - Z	1,7
Total damages & losses	= H	16,1			
Benefit of slash-and-burn 2)			**Cost of abatement & enforcement (A&E) 4)**	
Plantation companies	34%	2,7	Make-up lost profits	= B - Z	1.7
Private, community, government lands	66%	1.5 - 5.3	Legal and enforcement		0.5 - 1.0
Total benefits	= B	4.2 - 8.0	Make-up lost income		1,5
			Total A&E costs	= C	3.7 - 4.2

1) World Bank (2016). Direct costs H° incl. damages to agriculture (estate & food crops), losses to food crops,
   health treatment & fire fighting costs
2) Purnomo and Shantiko (2015), Rusli simplistically assumes same oil palm benefits for timber plantations
   (Private, community, government lands: min. if own use versus max. if on-sale to companies)
3) Gouyon and Simorangkir (2002), Rusli assumes profits for 2.6mn ha from burning and lower crop protection,
   neglect higher medium-term fertilizer cost. Rusli estimates profits for 37% peat swamps and 63% heavy forests
   (17% for timber, 46% for oil palm plantations)
4) Rusli estimates for legal enforcement and policing, assume full make-up of company profits and farmer incomes.
   Cost estimates exclude benefits of lower losses to estate & food crops and to forestry lands from zero-burning

Figure 2.   Cost Benefit Analysis

their medium-term relevance.[8] The aggregate benefits of slash-and-burn practices for plantation companies as well as farmers, residents and local elites is estimated to range from 4 billion US dollars to 8 billion US dollars. Here we use published estimates, whereas for private, community and government (i.e., non-plantation and forestry company) lands we assume either own use or subsequent sale of the burned and cleared lands to the larger companies.[9]

Next we designate C as the aggregate costs of fire and haze abatement and enforcement to be incurred by either government, which includes any incentives or profit and income compensation to be offered to farmers and companies, as well as costs incurred to enforce and punish violators of zero-burning regulation. We estimate these aggregate abatement and enforcement costs to reach almost 4 billion US dollars. Initially we make the strong assumptions that the aggregate abatement and enforcement costs,

---

[8] See Gouyon and Simorangkir (2002).
[9] We interprete per hectare benefit estimates from Purnomo and Shantiko (2015).

and the effectiveness of these abatement and enforcement efforts to induce zero-burning, are the same regardless of whether these are implemented by the Indonesian or the neighboring government. Moreover we assume, plausibly, that such abatement and enforcement costs are lower than the direct and indirect costs of allowing forest fires and slash-and-burn techniques to continue unabated.

We define H as the aggregate direct and indirect damages and losses from forest burning and the resulting haze pollution. This comprises the aggregate costs for the Indonesian economy and government, H^i, which is estimated by the World Bank (2016) to reach 16 billion US dollars, as well as the aggregate costs to the neighboring economies, H^n, so that $H = H^i + H^n$. The neighboring countries' direct healthcare expenditures and losses to industry and tourism are harder to quantify. Nevertheless, previous numbers from the late 1990s estimated such costs in the order of magnitude of up to 2 billion US dollars, which today, given population and economic growth as well as inflation, would be multiples higher.[10] We thus plausibly assume

$$H > H^i > H^n > C. \tag{2}$$

Note that the estimated aggregate damages and losses include not only direct costs but also indirect losses. Direct costs cover the costs of firefighting and healthcare expenditures as well as damages and losses to agricultural and forestry lands and equipment. Indirect costs include losses to the regional economy including the transport, industry and tourism, environmental and biodiversity losses as well as indirect loss in wages due to fire- and haze-related sicknesses.

On the basis of this simple framework with high aggregate direct and indirect costs, damages and losses and modest abatement and enforcement costs, the payoff structure of this game, assuming the two players bargain simultaneously, looks as follows:

When neither government chooses to invest in abatement and enforcement, *[Indonesia, Neighbor]* = *[Not-invest, Not-invest]*, the forest fires

---

[10]Taccioni (2003) summarizes the Asian Development Bank (ADB) and Institute of Southeast Asian Studies (ISEAS) estimates, whereby the ADB indicates that the indirect environmental and carbon emission losses to neighbouring countries i.e., Malaysia and Singapore make up the largest proportion of the neighboring economies' losses from the Indonesian haze.

and haze remain unchecked and the payoffs are $[B - H^i, - H^n]$. Indonesia keeps the benefits of forest burning but incurs heavy costs, damages and losses to its citizens, its economy and the environment, while the neighboring government suffers the health and economic costs of the transboundary haze. By contrast, if the Indonesian government decides to unilaterally invest in abatement and enforcement, and is effective at it, *[Invest, Not-invest]*, it reduces its domestic benefits and incurs the aggregate costs of abatement and enforcement, yet avoids the costs of damages and losses from the fires and haze. The payoff shall be $[Z - C, 0]$ where the neighboring country enjoys a haze-free sky without incurring the costs, apart from low cost technical cooperation and political coordination efforts (which we ignore). Should the neighbouring government decide to unilaterally invest, *[Not-invest, Invest]*, however, the payoff shall be $[Z, - C]$. Last but not least, if both governments decide to jointly invest and share the costs of abatement and enforcement, the payoffs change to $[Z - (1 - \gamma)C, -\gamma C]$, where γ is the fraction of abatement and enforcement costs that the neighboring government bears if the Indonesian government was prepared to cooperate and jointly invest.

Now applying the relative benefits (1) and costs (2) to the relative estimates in Figure 2, we can infer the relative payoffs

$$(Z - C) > (B - H^i) \text{ and } Z > (Z - (1 - \gamma)C) \qquad (3)$$

for the Indonesian government, and

$$- C > -H^n \text{ and } 0 > - \gamma C \qquad (4)$$

for the neighboring country government. Here the Indonesian government could be incentivized to proactively abate and punish slash-and-burn practices across the country if it believes that the neighboring country government will not invest in abatement and enforcement. However, if it thinks that the neighboring government will invest in abatement and enforcement, the Indonesian government is incentivized not to invest, and will maintain status quo. On the other hand, the neighbouring country government is only incentivized or forced to invest in abatement and enforcement if the Indonesian government is expected to not invest; it will not invest if the Indonesian government shows any willingness to invest. This is a classic "game-of-chicken" with three Nash equilibria, two pure- and one mixed strategy. The first pure one, *[Not-invest, Invest]* results in payoffs $[Z, -C]$ as per Figure 1, and represents the equilibrium where the Indonesian government will not invest if the neighboring government is willing to invest. The second Nash equilibrium, *[Invest, Not-invest]* with payoffs $[Z - C, 0]$ describes the outcome where the Indonesian government will invest if the Singaporean government credibly commits to not investing at all. The third Nash equilibrium is a mixed strategy, where each player chooses Invest or Not-invest with a certain probability.

Analysis of the haze game

Before we move on to analyzing possible solutions strategies to this haze game-of-chicken, a few points are noteworthy. Firstly, we examine the assumption of comparable effectiveness of abatement and enforcement efforts. For example, the neighboring government's investment in the full abatement and enforcement costs may not be effective in completely eradicating slash-and-burn practices. Possible reasons may be resistance and insufficient cooperation from local governments and business elites in the relevant regions in Indonesia, inadequate data on concession bounda-ries and fire locations, as well as insufficient knowledge of the local bureaucrats' and business elites' strategies and incentives. As a result, residual damages and losses $H^r = H^{ri} + H^{rn}$ may become unavoidable even after the neighboring government has spent the full amount C. Making the plausible assumption that $H^{ri} < H^i$ and $H^{rn} < H^n$, this would lead to a different payoff structure of the haze game. In particular, the payoff of the first Nash equilibrium *[Not-invest, Invest]* may shift to $[Z - H^{ri}, -C - H^{rn}]$. If we assume that both governments agree to share the costs of abatement and enforcement, the Indonesian government would now be prepared to co-invest provided it does not have to share too much of the abatement and enforcement costs, i.e., when

$$(Z - H^{ri}) < (Z - (1 - \gamma)C). \tag{5}$$

In this case, investing in abatement and enforcement programs now becomes the dominant strategy for the Indonesian government. Knowing that the Indonesian government will invest, the neighboring government now finds it advantageous to free-ride and not invest at all. Thus as soon as the residual damages and losses approach or exceed the cost of abatement and enforcement, $H^{ri} \rightarrow C$, the game may lead to one single Nash equilibrium, *[Invest, Not-invest]*. In other words, the neighboring government may not put in optimal efforts or, shirk together.

Note that under the original assumption of effective abatement and enforcement by either government, the aggregate payoffs of the two Nash equilibria, and the fully cooperative outcome of joint investing, achieve the same at $Z - C$. However, if the neighboring government is less effective in enforcing zero-burning policies in Indonesia, the resulting unique Nash equilibrium of *[Invest, Not-invest]* is the globally welfare-optimal solution. As we shall elaborate later, policy prescriptions should then be identified to nudge the game into this second pure Nash equilibrium.

Secondly, the Indonesian government, whether due to bounded rationality, local political and business interaction or myopic short-term political considerations, may not directly observe or may not take into account the indirect environmental costs, productivity and industrial losses. From the Cost Benefit Assessment in Figure 2, one could assume that the costs of fire and haze observed and considered by the government are limited to a subset H^{i0} of the direct damages and losses only, $H^{i0} \ll H.$[11] Concurrently, let us assume that the local political economy dynamics in the relevant regions may render it more difficult and costly for the government to satisfy the incentive constraints of the local bureaucrats, business elites and farmers, rendering the abatement and enforcement costs sufficiently high, so that $C > H^{i0}$. This scenario is possible when the Indonesian government, or a subset of it that may include local bureaucrats, central legislators and business interests, is politically influential and can drive or impede the central government's negotiations, and is concerned only about the direct health care expenditures and damages to equipment and infrastructure. In this case, the payoff structure changes such that for the relevant or powerful part of the Indonesian government

---

[11] See the next sub-section on common agency and local capture. From Figure 2, health and firefighting costs amount to 400 million US dollars, while direct damages from burned agriculture and timber lands, which have typically already had prior utilization, can be much lower than the total agricultural and forestry costs (as discussed by World Bank (2016)).

Not-invest becomes the dominant strategy because $B - H^{i0} > Z - C$ and $Z > Z - (1 - \gamma)C$. This in turn allows it to force the neighboring government to invest in abatement and enforcement because $-C > -H^n$; it is now costlier for the neighboring government to leave the fire and haze unabated, than to invest in the necessary measures. The new Nash equilibrium ends up at *[Not-invest, Invest]* and the Indonesian government is now happy to free-ride on its neighbor. Note, however, that this only holds as long as the residual direct health and damages cost to Indonesia, which may be non-zero if the neighboring government may not be as effective in implementing the abatement and enforcement measures, is not too large such as to violate $Z - H^{i0r} > Z - (1 - \gamma)C$, where we now define H^{i0r} as the subset of direct health and damages cost to the relevant parts of the Indonesian government. If $Z - H^{i0r}$ does fall below $Z - (1 - \gamma)C$, both governments would have to accept the benefits of cooperating and jointly investing in abatement and enforcement.

Thirdly, it should be mentioned that in this simplified game framework we neglect the negative domestic and international reputational and political costs of the Indonesian government not proactively making efforts and investing in fire and haze abatement and enforcement. These costs are more difficult to quantify. Moreover, the abovementioned myopia and political pressure, combined with agency and capture problems, may force the government to ignore or underplay the reputational issues of not investing. In any case, it will try to introduce measures that may give the appearance of it trying hard to mitigate the forest fires and incentivize the various parties to adhere to the zero-burning laws and regulations.

Common-agency and local capture

Another important aspect of this Indonesia haze game framework concerns the incentives of the key players involved in this game. While a rational and benevolent central government in Indonesia may look at the aggregate costs and benefits of fire and haze abatement and enforcement, it has become clear that the central government has had problems aligning the interests of national legislators, local governments and legislatures, business elites, farmers and residents. Observations made by Nguitragool (2011) and Purnomo and Shantiko (2015) help inform of the complex local common-agency problem faced by the Indonesian government.

Nguitragool summarizes the problems faced by the central government, which led Indonesia's negotiations with its neighbours on the ASEAN Haze Treaty. These problems centered around the environmental ministry's lack of authority, the increased autonomy of local district-level governments and the delayed ratification of the Haze Treaty by the legislature. Purnomo and Shantiko are able to identify the key beneficiaries of slash-and-burn practices, which together enjoyed roughly 3,077 US dollars per hectare of benefits that included plantation companies (around 30 percent of total benefits), various groups of small and individual farmers and residents involved in slashing, cutting and burning (14 percent), village heads (3 percent) as well as, in particular, local group organizers which primarily comprise local business elites.

As can be inferred from the simplified scheme in Figure 3, each of these parties enjoy distinct benefits and are affected in different ways by the forest fires and haze pollution. Local farmers and residents, in particular, are influenced by multiple principals, including business elites, plantation companies, as well as local bureaucrats. Business elites with close connections to local governments and legislatures as well as local and national plantation companies may spot profitable opportunities such as slash-and-burn land clearing projects. They usually target the smaller local and national companies and groups of exempted small farmers who will benefit most from the cost advantages of slash-and-burn while being less exposed in terms of reputational risks. These business elites may also be able to capture and convince local bureaucrats to either look away or, in some instances, help hide and manipulate data about burning acreage and their origins. The local bureaucrats could be tempted to collude with the local business elites, and may seek the latter's help in convincing the local legislature members not to object to these potentially lucrative business dealings, and to persuade their national-level party colleagues to resist onerous laws and regulations that may jeopardize such business activities.

It is noteworthy that the more lucrative business transactions tend to generate higher income to the local economy, resulting in higher local tax receipts as well. On national- and international-level, larger plantation or forestry companies may decide that to invest in or purchase cleared and burned lands may be too big of a reputational and business risk for them. Instead they may opt to assist the government in its abatement and enforcement efforts, and employ the farmers and residents to aid in their zero-burning mechanical land clearing works.

Through all this, local farmers and residents especially in less-industrialized regions, who do not have tangible income alternatives other than working in plantation acreage owned by themselves or by the larger companies, have to choose between engaging in the often more lucrative burning activities and the wages they might earn from working for larger companies conducting legitimate activities only. While individual and small groups of farmers with combined acreage of less than 20 hectares are by law permitted to engage in controlled forest burning, it is usually the business elites that are proficient in coordinating individual and small farmer groups and bundling and on-selling such transactions to plantation companies that are willing to take the risk.

The above analysis demonstrates that the overall fire and haze negotiation process is not led by the Indonesian central government as one coherent authority, but is affected by multiple decision makers operating in a complex multi-principal common-agency environment, each with their own cost-benefit calculations. When one adds to this the possible myopia and bounded rationality problems faced by the various players, the above

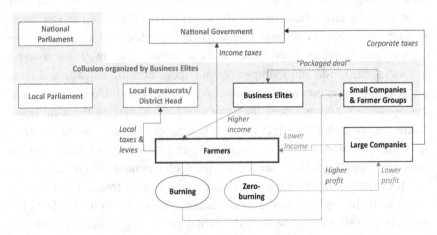

Figure 3.   Local Political-Economy Scheme

simplified game-of-chicken becomes a more complex multi-principal multiple-task bargaining process. For example, just like the purely utilitarian business elites, local bureaucrats, farmers and residents are less likely to consider the broader, indirect economic and environmental losses of fire and haze. At the same time, plantation companies, except for the larger national and international ones, tend to focus on their profit maximization objective, with lesser regard for reputational and long-term costs.

Only the central government and environmental ministry are able to and may consider all indirect damages and losses of fire and haze in their decision making and negotiation arguments. However, they may not observe all indirect environmental, biodiversity and illness-induced wage losses. In fact the central government and its environmental ministry face a complex adverse selection and moral hazard problem. Firstly, it may not be able to identify exactly which farmers, companies and government bureaucrats and bodies are involved in the illegal slash-and-burn activities, who may as a consequence try to impede investigations and the passing of preventive and remedial legislation (an adverse selection challenge). Secondly, central government may not observe the true abatement and enforcement efforts made by local bureaucrats as well as companies who may have been captured by the business elites (a moral hazard problem). Paralleling Nguitragool (2011), it is our view that it is these domestic institutional and governance challenges, as well as incentive incompatibility problems, that have significantly delayed the ratification of the ASEAN Haze Treaty and have rendered the coordinated abatement enforcement efforts of Indonesia and its neighbors difficult and protracted.

Solution strategies and policies

There are several solution scenarios for the Indonesian haze game-of-chicken. First, note that a classical simultaneous game-of-chicken with two pure-strategy Nash equilibria, *[Invest, Not-invest]* and *[Not-invest, Invest]*, also finds a third unique mixed-strategy equilibrium where each of the two players credibly conveys their strategy of Not-invest or Invest with fixed yet distinct probabilities. The mixed strategy probability chosen by a player will ensure that the expected payoff of the counter-party is equalized regardless of whether the counter-party chooses to invest or not invest. Were we to view this Indonesia haze game as an annually recurring repeated chicken game over a finite number of time periods, theory tells us that there exist many subgame perfect equilibria. In practice, however, one could think of this as a sequential game, which possesses two symmetrical Nash equilibria of *[Not-invest, Invest]* or *[Invest, Not-invest]* whenever the first mover credibly commits to choosing either not to invest or to invest, and the second mover reacts rationally by choosing to invest or respectively not to invest.

If the Indonesian government happens to be the first mover that credibly signals that it will not invest at any cost, the solution will be *[Not-invest, Invest]*. This Nash outcome is then the same as the more realistic case where the Indonesian government is dictated by political and business pressure to only consider the direct costs of fire and haze and is thus incentivized to free-ride on the neighbouring government.[12] Such a sequential game with a unique Nash equilibrium, if repeated for a finite number of time periods and played under conditions of complete information, will by rational backward induction have a unique subgame perfect equilibrium as well, which is *[Not-invest, Invest]*. As long as Indonesia credibly signals its willingness to tolerate the negative consequences of forest fires and pollution, or its inability to enforce zero-burning, it will force the neighboring government to invest in abatement and enforcement.

Under incomplete information, however, the subgame perfect equilibrium will depend on the first mover's belief of the second mover's reputation and possible decision. On the one hand, if the first mover does not know for sure of the second mover's possible reaction and thus believes that with a certain probability the second mover may, for the sake of playing hardball, insist on not investing as well, theory shows that for a game played sufficiently long periods of time the first mover may end up choosing invest every time. The subgame perfect equilibrium then becomes *[Invest, Not-invest]*. On the other hand, if the first mover believes that the second mover is more likely to relent, it may choose to not-invest, resulting in the equilibrium *[Not-invest, Invest]*.

---

[12] Recall the chain store paradox introduced by Selten (1978) and analyzed by Kreps *et al.* (1982) and many others later on.

A second approach to solving this game-of-chicken is to change the payoff structure of the game. For example, neighboring governments can force the Indonesian central and local governments and legislatures to consider the total aggregate direct and indirect costs H^i, while reducing the cost of abatement and enforcement C by improving the effectiveness of transparent forest monitoring and zero-burning enforcement. Especially in an environment where the neighboring government may have difficulties in fully abating and preventing slash-and-burn practices, $H^{ir} > 0$, the neighboring government could force Indonesia to consider investing in abatement and enforcement as a dominant strategy, unilaterally or jointly with the neighbouring government. Another tactical move could be for the neighboring government and the international community to raise the stakes and the reputational risk of the Indonesian government not investing in abatement and enforcement.

To force the Indonesian government and negotiators to gain the support from all relevant local constituencies, however, requires a third strategy that can address the local-level common-agency and capture problems. It is noteworthy that in a multi-principal and multi-task context such as the Indonesian haze game, where one task such as slash-and-burn is highly profitable for the colluding principals (i.e., the local bureaucrats and business elites), and thus as a consequence, *ceteris paribus*, also for the farmers involved in such slash-and-burn activities, it is more cost-effective to concurrently ensure severe punishment and strict enforcement of laws and regulations, than to solely incentivize zero-burning with financial transfers. The reason is that solely incentivizing the alternative task (for companies, adherence to zero-burning and for farmers, earning legal wages for mechanical land clearing) will result in very high costs to the central government if the latter would have to financially incentivize the farmers to adhere to zero-burning (Dixit, 1996).[13] This approach is consistent with Holstrom and Milgrom (1991), who find that if the outcome of one task is poorly observable (i.e., which individual farmers and companies do adhere to zero-burning), then incentive for the more observable task (of slash-and-burn) must be reduced, or in our context, strictly prohibited.

---

[13] See the Appendix of Dixit (1996) for a multi-principal and multi-task model.

Additionally, we could also adopt the solutions to the multi-principal common-agency model of Bernheim and Whinston (1986), where again the effort of the agent (i.e., the farmers) is unobservable, but the outcome (burned lands, or fewer burnings) is observable. Here multiple principals gain from cooperation if they possess similar information. Therefore, if the colluding principals' coalition can be broken, for example by creating information asymmetry within the group of colluding principals, and if the agents are risk averse, the latter's incentives to continue engaging in slash-and-burn practices become weaker. One way to break such a coalition of local principals is to prevent each principal from observing each other's outcomes; in fact, in the extreme case where the agent's efforts on behalf of different principals' become perfect substitutes, the equilibrium actually becomes first best. Following Tirole (1994), furthermore, the compartmentalization of responsibilities across the different local principals would work in the central government's nation-wide welfare-maximizing interest.[14]

We shall now discuss how these haze game and local agency and capture considerations can help neighboring governments identify welfare-enhancing solution strategies and policy options for the ASEAN region, which they can jointly implement and invest in together with the Indonesian central and local governments.

Changing the payoff structure of the game

Policies must be identified that incentivize the Indonesian government as well as the powerful local players to neither free-ride on its neighbours nor test the neighboring government's resolve to play hard ball. Such policies require that neighboring governments introduce policies that shape the Indonesian counter-parties' beliefs of the neighboring government's type and possible actions, and change the payoff structure of the game for each of the relevant and powerful local players. The optimal policy package should include, firstly, educating about and increasing the direct and

---

[14]Many other authors, e.g., Martimort (1996), Grossman and Helpmann (1994), Epstein and O'Halloran (1996), find that more intense inter-principal competition will allow higher efficiency.

indirect costs of haze and pollution, and incentivizing the local bureaucrats to take into account nation-wide interests in their decision making (increase $H^{i0} \rightarrow H^i$). The neighbouring government must continue warning residents, farmers and local bureaucrats about the health hazards and, equally importantly, educating them of the indirect environmental and productivity losses from the haze and pollution. It should offer technical assistance and collaborate with the Indonesian central and local governments to help introduce and implement effective surveillance and database technologies to improve transparency and awareness of the aggregate costs of fire and haze. Secondly, both governments must work together to identify ways to reduce the cost of abatement and enforcement (reduce C). These include improved surveillance, systematic and accurate data collection and processing to aid early warning systems about fire locations, as well as intensive monitoring of local bureaucrats and business elites. Thirdly, the benefits of and profits from slash-and-burn practices should be reduced through strict enforcement of zero-burning laws and regulations, as well as severe punishment both locally, nationally and, for large international companies, through their business and financial partners in the neighboring countries (decrease $B \rightarrow Z$). For instance, prosecution and taxation of Indonesian and neighboring country-based plantation companies provide the neighbouring government(s) the means to actively deter slash-and-burn practices.[15] Fourthly, the central government should incentivize and subsidize legitimate zero-burning land clearing practices, (to increase $Z \rightarrow B$), as well as to encourage and support alternative employment possibilities. Education and training programs inform residents and farmers, while appropriate industrial strategies open alternative employment and business opportunities for local farmers and residents. Transfer mechanisms need to be identified that can help compensate for local farmers' lower income and to incentivize local bureaucrats to adhere to and help enforce zero-burning practices.

The rationale of this policy mix can again be inferred from the normal-form game structure in Figure 1. The higher $H^{i0} \rightarrow H^i$ and the smaller the difference $(B - Z) \rightarrow 0$, the more likely it is that the policy to invest becomes the dominant strategy for the Indonesian government,

---

[15] See Tan's (2015) analysis of Singapore's Transboundary Haze Pollution Act.

resulting in a single Nash equilibrium of *[Invest, Not-invest]*. Following this, the haze stage game can be expanded to its practical nature of a repeated, sequential game-of-chicken. While offering technical support to the Indonesian central and local governments, the neighboring government can approach the negotiations in one of two ways. On the one hand, it could play hardball and credibly signal its readiness to not-invest in abatement and enforcement, even at the expense of tolerating the seasonal haze pollution problem. This forces the Indonesian government, assuming the aggregate abatement and enforcement costs are low enough and the aggregate damages and losses high enough and observable to the relevant parties, to unilaterally invest in abatement and enforcement. On the other hand, the neighboring government's reputation and past actions may indicate that it is not ready to tolerate sustained haze pollution; thus its insistence on playing hardball and to strictly not-invest is not credible. In this case, the neighboring government could offer to share part of the costs of abatement and enforcement, just enough to catalyze and induce the Indonesian government to invest. Note that *[Co-invest, Co-invest]* is not a Nash equilibrium of the game. Nevertheless, a complex interplay of historical path dependence, reputation and incomplete information about the indirect costs of fire and haze may force the parties to compromise. Thus the outcome of the negotiations may resemble the above mentioned mixed-strategy equilibrium outcome.

Aligning local incentives

Several of the abovementioned policy options also help mitigate the local multiple-principal and common-agency problems. In particular, the Indonesian and neighboring government should collaborate in breaking the collusion between local bureaucrats and business elites. This can be done by concerted local public relations programs and by empowering local and national anti-corruption institutions. At the same time, local bureaucrats can be incentivized by supporting and subsidizing them with important and visible infrastructure projects, which will help improve their standing in the local communities.

To enhance the collection of accurate fire data, national and local regulatory institutions and non-governmental organizations (NGOs) should collaborate with local bureaucrats in monitoring the behavior of plantation companies. Concurrently, however, national regulators and NGOs should use satellite imaging and other surveillance technologies to collect accurate data independently from the local bureaucrats and their local appointees.

As mentioned above, residents and farmers should be offered training and education that fit the national and local government's industrialization program, offering them alternative employment and income opportunities.

Summary

We identify solution strategies and propose policies to abate man-made forest fires in Indonesia, mitigate the resulting transboundary haze pollution and enforce zero-burning regulation. Several policy measures should be prioritized. Firstly, residents, farmers and local bureaucrats should be educated about the increasing direct and indirect costs of haze and pollution. Secondly, local bureaucrats should be incentivized to take into account nation-wide interests in their decision making. Thirdly, both central and local governments must work together to identify ways to reduce the cost of abatement and enforcement. Fourthly, the benefits of and profits from slash-and-burn practices should be reduced. These can be achieved through strict enforcement of zero-burning regulations as well as severe punishment. The latter must be introduced on both local and national levels, and for large international companies, through their business and financial partners in the neighbouring countries. Fifthly, the central government should incentivize and subsidize legitimate zero-burning land clearing practices as well as encourage and support alternative employment possibilities. Throughout all this, neighboring country governments can and should offer technical and financial incentives to the Indonesian central and local governments, residents and farmers.

References

Adriani, M. (2016). *Methodology in Estimating Damage and Losses of Indonesia Forest Fires in 2015*, World Bank: Jakarta, Indonesia.

Bernheim, B. D. and Whinston, M. (1986). Common agency, *Econometrica*, 54(4): pp. 911–930.

Bertinelli, L., Marchiori, L., Tabakovic, A., and Zou, B. (2017). The impact of unilateral commitment on transboundary pollution. *Environmental Modelling & Assessment*, pp. 1–13.

Carraro, C. and Sinisalco, D. (1995). Policy coordination for sustainability: Commitments, transfers, and linked negotiations, in Goldin, I. and Winters, L. A. (eds.), *The Economics of Sustainable Development*, Centre for Economic Policy Research, pp. 264–288.

Chander, P. and Tulkens, H. (2006). Cooperation, stability and self-enforcement in international environmental agreements: A conceptual discussion. FEEM Working Paper No. 34.

Dixit, A. (1996). *The Making of Economic Policy: A Transaction-Cost Politics Perspective*, The MIT Press: Cambridge, MA.

Duek, A., Brodjonegoro, B. and Rusli, R. D. (2010). Reinterpreting social processes: How system theory can help to understand organizations and the example of Indonesia's decentralization, *Emergence: Complexity & Organization*, 12(4): pp. 30–56.

Epstein, D. and O'Halloran, S. (1996). A theory of strategic oversight: Congress, lobbyists, and the bureaucracy, *Journal of Law, Economics and Organization*, 11: pp. 227–255.

Fünfgelt , J. and Schulze G. G. (2011). Endogenous environmental policy when pollution is transboundary, University of Lueneburg Working Paper Series in Economics Vol. 196.

Gouyon, A. and Simorangkir D. (2002). *The Economics of Fire Use in Agriculture and Forestry: A Preliminary Review for Indonesia*, WWF: Gland, Switzerland.

Grossman, G. M. and Helpman E. (1994). Protection for sale, *American Economic Review*, 84(4): pp. 833–850.

Holmstrom, B. and Milgrom P. R. (1991). Multitask principal-agent analysis: Incentive contracts, asset ownership, and job design, *Journal of Law, Economics and Organization*, 7: pp. 24–51.

Kreps, D. M., Milgrom, P. R., Roberts, J. and Wilson, R. J. (1982). Rational cooperation in the finitely repeated prisoners' dilemma, *Journal of Economic Theory*, 27(2): pp. 245–252.

Martimort, D. (1996). Exclusive dealing, common agency, and multiprincipals incentive theory, *The RAND Journal of Economics*, 27(1): pp. 1–31.

Nguitragool, P. (2011). Negotiating the haze treaty: Rationality and institutions in the negotiations for the ASEAN Agreement on Transboundary Haze Pollution 2002, *Asian Survey*, 5(2): 356–378.

Purnomo, H. and Shantiko B. (2015). The political economy of fire and haze: Root causes, Discussion Forum: Long-term Solutions to Fires in Indonesia:

Multi-Stakeholder Efforts and the Role of the Private Sector, Global Landscape Forum: Paris.

Selten, R. (1978). The Chain-Store Paradox, *Theory and Decision*, 9(2): pp. 127–159.

Simorangkir, D. (2007). Fire use: Is it really the cheaper land preparation method for large-scale plantations? *Mitigation and Adaptation Strategies for Global Change*, 12: pp. 147–164.

Tacconi, L. (2003). Fires in Indonesia: Causes, costs and policy implications, CIFOR Occasional Paper No. 38.

Tan, A. K.-J. (2015). The haze crisis in Southeast Asia: Assessing Singapore's Transboundary Haze Pollution Act 2014, NUS Law Working Paper 2015/002: Singapore.

Tirole, J. (1994). The internal organization of government. *Oxford Economics Papers*, 46: pp. 1–29.

World Bank. (2016). *The Cost of Fire: An Economic Analysis of Indonesia's 2015 Fire Crisis*, World Bank: Jakarta, Indonesia.

17 The Southeast Asian Haze: Dealing with an Outcome of Global Climate Cycles and Change

Victor R. Savage

*Visiting Senior Fellow, S. Rajaratnam School of International Studies,
Nanyang Technological University, Singapore
isvrsavage@ntu.edu.sg*

Introduction

Nature by and large changes ecologically in an orderly and predictable manner — the four seasons, the monsoon seasons, and the mating seasons for many insects and mammals. The Western world saw this regularity of nature as part of heavenly order God created — which often underscores the idea of the 'design earth' (Glacken, 1967). Yet amidst this seemingly peaceful change in temporal terms, Nature also can unleash its violent, sudden and devastating outbursts in volcanic eruptions, earthquakes, tsunamis, and floods with little prior warning. Human and ecological systems are mainly the victims of the power of Nature's wrath. Given changing climatic conditions in the past, some scientists wonder whether climate change is a product of the Anthropocene or an independent aspect of changing nature (Bender, 2013). All aspects of Nature's change however can be seen both positively and negatively by human societies. Volcanic eruptions and floods are for many agrarian societies a positive feature in providing rich fertilizers. For centuries, volcanoes and mountains were the homes of Gods and spirits and continue to be worshipped by communities in Southeast Asia. However for urban communities in Thailand, floods are a bane to living, disrupting transport systems, undermining industrial plants and creating damage to homes. The mega floods of 2011 in Thailand are a case in point. It killed over 500 people, affected the lives of 13.6 million people, and left economic damage of 46 billion US dollars (https://en.wikipedia.org/wiki/2011_Thailand_floods. Accessed June 1, 2017)

Amidst these contrasting scenarios of a changing and temperamental Nature, lies issues such as forest fires and their resulting transboundary haze which continually plagues many countries in Southeast Asia. How do we define these issues? Are they products of Nature that human societies need to adjust to? Or are they products of human activities which can be curbed with political action? Increasingly, scientists, academics and the informed public are beginning to realise that human beings are changing the face of the Earth. We more than any force of nature are the shapers and modifiers of nature and environmental change. Flannery (2006) underscored the climate change debate when he stated that human beings are now the 'weather makers'.

We do not know the exact scientific processes of these changes fully but we certainly are experiencing all sorts of freak weather conditions and environmental changes — global warming, heat waves, floods, intense rainfall, prolonged droughts, sea level rises, coral bleaching, mass biodiversity extermination, increasing frequencies of typhoons and cyclones which all lead to other cascading ramifications on food and water security, human health, malnutrition, unemployment, poverty, rural debt, social tensions, increasing disparities of wealth, urban riots and the undermining of development in lesser developed countries.

Climate change questions lie at the crux of the issue of the environmental haze in Southeast Asia which have resulted from massive forest fires in Kalimantan Borneo and Sumatra and affected Malaysia, Brunei, Singapore, the Philippines and Thailand in 2015. In mainland Southeast Asia, Thailand is the major culprit of fires and haze due to farmers burning their corn farms to reuse them for planting more corn for animal feeds. Between 2008 and 2015, maize plantations in Thailand expanded 77,880 hectares and yet the harvest of 4.6 million tons of corn in 2015 could not meet local demands (Tan, 2016: A11). So long as the forest fires and haze fall within countries they have to be tackled by national governments; but when such fires create transboundary haze they become issues of international and regional concern and hence underscore the politics of climate change (Giddens, 2009).

The complexity of the fires and resulting haze problem provide no easy answer and solution to varied governments faced with this environmental problem. This is an environmental challenge which has political, economic, social and ecological underpinnings. The issue of clearing forests might be an economic benefit for Indonesia wrestling with poverty, inequality and development challenges but for its neighbors it is an irritant undermining the quality of living and health. Any regional solution involves sovereignty issues which need sensitive handling as the Singapore government has unpleasantly found out with regard to its legal proceedings against companies involved in clearing of land via burning. In Indonesia, palm oil conglomerates are heavily involved in the political landscape and singling any company out for errant corporate behavior is not easy without affecting political sensibilities.

This paper argues that while the genesis of Southeast Asian forest fires and the haze are embedded in much wider global climatic changes, one cannot deny that human beings are culprits of forest fires and the concomitant haze. The solutions to forest fires and the haze require both mitigating and adaptive responses. Some aspects of adaptation such as staying indoors in air-conditioned offices and spaces to reduce haze is often misplaced. The haze permeates even indoor air-conditioned spaces because air-conditioners take in fresh air which means during the haze the indoor air-conditioned air is not haze free. The issue is not national, regional but global as well as there are no easy solutions to mitigate forest fires and haze. So long as El Niño and climate change are trigger mechanisms for forest fires and resulting haze, no government or consortium of governments can solve this challenge until we reduce CO_2 emissions globally. Ironically it is because Southeast Asian states are economically linked globally, the economic propensity for cheap fire clearings of forests and agricultural waste becomes a worthwhile economic proposition. Similarly the globalizing impact of CO_2 increases resulting in global warming and prolonged dry weather conditions become a viable economic proposition for local companies, multinational corporations (MNCs) and even governments to abet cheap land clearance from fires. In short, global warming is a surrogate for forest fires and its resulting haze. Thus, the wider the scale of the issue the more complex it becomes to solve it.

Contextualizing the 'hazy' challenge

For over a century of recorded history, the region has faced rather extreme weather conditions. As science has progressed, the international community is getting a better understanding of the interconnections of climatic issues on a global scale. In many of the ecological issues facing specific areas, the tendency in the past is to look *in situ* at variables to explain why certain catastrophes take place. In some environmental cases, the *in situ* investigations do hold up such as the propensity for landslides and mudslides to frequently reoccur in the same area.

In the case of the extreme dry weather conditions which frequently take place in Southeast Asia, there is no direct correlation of *in situ* environmental factors which lead to the drying and drought phenomenon problem in the region. What has been found by climate scientists is the discovery of the *El Niño* and *La Nina* phenomenon in the Pacific Ocean off Peru which has a direct correlation with devastating drought in mainland and insular Southeast Asia. *El Niño* is a product of the heating of the equatorial Pacific due to the weakening of trade winds which would usually push sun-warmed oceanic waters west from Latin America. When the warm waters linger, it triggers a reaction in the atmosphere. *La Nina* on the other hand is triggered by cold waters off Peru rising to the surface and causing worldwide heavy precipitation and drought. Fortunately, the *El Niño* and *La Nina* induced climate phenomenon only takes place once in several years and hence the region is spared the devastating impacts. The scientific community is slowly beginning to unravel the complex interrelationships between the ocean and atmosphere. But currently, William Rosen's (2014: p. 9) observations remain pertinent: "weather and climate remain the product of complex interactions between ocean and atmosphere, a dance set to almost unimaginably complicated rhythms, made even more complicated because one partner — the atmosphere — is enormously quicker to respond to change than the other."

Climate change questions lie at the crux of the issue of the environmental haze in Southeast Asia which have resulted from massive forest fires in Kalimantan and Sumatra and affected Malaysia, Brunei, Singapore, the Philippines and Thailand in 2015. In mainland Southeast Asia, Thailand is the major culprit of fires and haze due to farmers burning their corn farms to reuse them for planting more corn for animal feeds. Between 2008 and 2015, maize plantations in Thailand expanded 77,880 hectares and yet the harvest of 4.6 million tonnes of corn in 2015 could not meet local demands (Tan, 2016: A11). So long as the forest fires and haze fall within countries they have to be tackled by national governments; but when such fires create transboundary haze they become issues of international and regional concern and underscore the politics of climate change (Giddens, 2009).

The August 2015 haze was so bad that visibility in Palembang (Sumatra) was down to only 300 metres. In the Riau province in September 2015 the Pollutant Standards Index (PSI) reached 436 — the highest level for 2015. The Indonesian forest fires and haze were critical that President Joko Widodo on September 7, 2016 charged the chief of the National Disaster Management Agency (BNPB) to tackle the complex issue. In Malaysia, 2,000 schools were closed which affected 1.5 million students. At the international front, Singapore's Defence Minister Ng Eng Hen met with his Indonesian counterparts in September 2015 to explore ways of cooperating to mitigate the fire and haze issue. Singapore's then Environment Minister, Dr Vivian Balakrishnan acknowledging Indonesia's efforts argued the need for "closer regional and international cooperation to apply legal and commercial pressure" (Sim, 2015: A8).

In many rain-fed agrarian economies, prolonged drought spells disaster for crops which in turn leads to famine and human deaths. Over the last century, *El Niño* has peaked in terms of its devastating impacts on human societies. For the Southeast Asian region in recent memory, the three strongest *El Niño* impacts took place in 1994, 1997–1998 and in 2015. Much has been written about the 1997–1978 forest fires and haze (Asian Development Bank, 2001) and there is no need to rehearse the arguments here. Unfortunately in 2015, *El Niño* took place in a year of extreme climate change — 2015 was the hottest year on record since temperatures have been recorded. And this trend is likely to continue. February 2016 was the hottest month in the world since the time weather records were taken. The impacts of weather changes on human societies in 2015 has been dramatic, painful and seemingly unexplainable. We have become victims and witnesses to Nature's wrath, or James Lovelock's view (2007) of Gaia's fever and others might say of God's anger.

Who is to blame: Nature or human beings?

Given the geopolitical nature of the haze affecting many countries and communities in a transnational manner, the ownership of what causes the haze becomes a hotly debated topic with regional ramifications. One must be reminded that the haze is only one form of air pollution which Asian countries and cities are subject to (Ohara *et al.*, 2005) but it is the most politicized because of its transnational implications. Indonesia is the source of the 2015–2016 forest fires that inevitably was the source of the regional haze. But the extended length and spatial span of the 2015 haze took most countries by surprise. Part of the extended fires is due to the fact they were uncontrollable because of the severe extended drought which made all vegetation easily consumed by fires. And secondly in Sumatra, the peatlands provide a natural energy reservoir for unstoppable fires — while fires might have died down peatlands keep the embers of fire burning which then flare up with winds. In the past 1997–1998 haze which was also equally bad, the impact tended to be intermittent — curiously the winds helped to spread the haze in different directions and the prolonged impact was somewhat lessened over time. However in the 2015 haze, the impact was more concentrated over long periods of time and hence the damage to economic activities and human health was more severe in Malaysia and Singapore.

Fire seems like a difficult element to define. It is rather rarely seen as a product of nature like water or earth. Yet it appears in two natural phenomena, in volcanic eruptions and through lightning. Amongst the Temiar in Peninsular Malaysia, lightning (a source of fire) coming from the sky is worshipped as God or the Temiar reference to *karey* (Benjamin, 2002). On the other hand, fire is also created by human beings for cooking, warming ourselves in winter and lighting a cigarette. So the nature of fire tends to be difficult to define clearly. Hindus use fire as a means of purification by washing their hands above fire just as they do with water. Forest fires are thus ambiguous phenomena. They can be sparked off in dry weather by winds creating friction in dry leaves. Or they can be set by human beings. These alternatives create confusion and open-ended blame on who is causing fires. This view that fires are natural accidents is held by Indonesian Vice-President Jusuf Kalla. He chastised President Susilo

Bambang Yudhoyono for apologising to Malaysia and Singapore about the haze. VP Kalla's reason was simple: "I have said before, this is a natural disaster, so there is no need to say sorry."

However the Indonesian administrators and corporate people have other culprits on who is causing Indonesia's fires. Among the causes are that swidden cultivators use fire to clear lands for planting which Clifford Geertz (1963) argues is a positive aspect of the agricultural ecology; so in the past plantation owners shelved the blame of forest fires to indigenous groups in Kalimantan in 1994. In the 1997–1998 massive forest fires, the cause of fires in Kalimantan and Sumatra was certainly not from indigenous cultivators but plantation owners. And the 2015 forest fires were certainly more a product of plantation companies seeking to clear lands rapidly, extensively and cheaply. From a macro perspective, the fires in Indonesia are products of both domestic and international blame.

Nature proposes and man disposes

Climatologists have increasingly been documenting the temporal phenomenon of *El Niño* induced drought in the region — some say it appears in seven-year intervals. In between, *La Nina* can also impact the region with an opposite impact of heavy precipitation leading to floods. Such scientific climatic data finally have come to the attention of historians, geographers and social scientists and hence history seems to be rewritten (Linden, 2007; Whyte, 2008). Suddenly climatic factors undergird major historical events and processes. A plethora of historical reinterpretations of civilizations across the world have appeared in recent years.

But despite the warming trends of the Holocene, two periods of intense cooling caused more damage in Europe that at any other time. In the 14th century, Rosen (2014) narrates the terrific catastrophe that engulfed Europe due to massive rainfall, floods and coldest winters that led to pestilence, warfare, animal epidemics and famine that claimed the lives of 6 million people. Europe experienced the 'Little Ice Age' in the 17th century which Parker's (2013) monumental work, *Global Crisis*, sums up best. The global cooling had profound political, social, cultural and economic impacts in Europe — leading to discord, upheavals, tyrannies, and changes of kingdoms (Parker, 2013: p. 11). There is no doubt that the combination of *El Niño* and climate change is likely to become socially disruptive and create political instability in the region in the coming decades if steps are not taken nationally and regionally to manage these issues.

In Southeast Asia, revisionist historians have finally found an underlying explanation to historical changes. Lieberman's (2003; 2009) two-volume historical study of mainland Southeast Asia was the first attempt by a historian to look at environmental and climatic impacts on historical processes and changes in the region. Not only did he ascribe the rise and demise of civilizations (Angkor Thom; Pagan) due to the *El Niño* and *La Nina* influences, he also took his climatic-historical thesis on a grand scale — linking similarities of historical processes in Europe, East Asia and Southeast Asia with what he called "strange parallels". In short, climatic changes (i.e., *El Niño* and *La Niña*) transcend other influences in explaining historical events worldwide. More than other factors, climate change is a global phenomenon. Climate change predates other human and social influences in what we define as 'globalization'. As historians are beginning to re-read history, climate change has become a new historical determinant as global catalysts and impactors on a grand spatial and global scale (Linden, 2007; Whyte, 2008; Parker, 2013; Rosen, 2014).

The issue of recurring *El Niño* and global climate change provides an important environmental understanding to forest fires and the haze in Southeast Asia. There is after all a metabolic interaction between human beings and nature which provides both an understanding of natural and socio-cultural human history. However Foster *et al.* (2010) in their thought provoking book, *Ecological Rift*, argue that human activities have created a 'metabolic rift' in human–nature interactions which underscore many of the current environmental and climate change problems. In the case of *El Niño*, revisionist historians are accepting its impacts in explaining global human historical events and processes — it endorses the adage that history repeats itself. The impact of *El Niño* in Southeast Asian history cannot be more instructive — the years of *El Niño* affected drought have led to civilizations collapsing and the years of good rains were periods of good crop harvest and expanding civilizations (Lieberman, 2003; 2009). Hence to understand Indonesia's forest fires and haze consequences during the *El Niño* years is documenting a cross-sectional event in current regional processes — with time, these events become part of historical interpretations in understanding economic, social and political issues both nationally and regionally.

El Niño is a cyclical component of Indonesia and insular Southeast Asia's history. All of us experiencing the forest fires and haze are ring side witnesses of human–nature relationships *in situ* (i.e., in specific place). The current 2015–2016 *El Niño* phenomenon is causing widespread havoc across the world in Southeast Asia, Australia, South America and India. These unfortunate events are the cause of human deaths, malnutrition, ill health (water- and mosquito-related diseases), and destruction of property as well as the extermination of countless millions of organisms. Armed now with science and technology one has to ask whether we can arrest the problems of fires, haze and destruction of all life, human and other biological specimens. The human impacts are a lot more widespread and alarming because of increases in population, the scale of cultural landscape change and the higher economic value.

While *El Niño* is a natural phenomenon in environmental history, global climate change seems a unique environmental product of the negative aspect of human–nature relationships. Climate change because of its human influence best defines the newly coined geologic time scale of the Anthropocene. We have become the shapers and engineers of the world ecology and landscapes. The eradication of forests in Indonesia whether by fire or other means is a major human influence resulting from global climate change and ecology. Forests are our carbon sinks, creators of life giving oxygen, and the source of our fresh water. When we exterminate forests, we are undermining our own human existence. Some 20 years ago, biologists Richard Leakey and Roger Lewin stated that the 21st century would be the age of the sixth great extinction (Flannery, 2015: p. 55) and this view has been given scientific substance by Elizabeth Kolbert (2014) in her book, *The Sixth Extinction.*

When Indonesians destroy their forests, they should ponder about the impacts on their own regional ecologies and national environments which will severely undermine all life and living. With the destruction of forest and environmental changes come increasing CO_2 emissions and global warming. As an archipelago, Indonesians are often reminded that hundreds of islands in their archipelagic state will be drowned out due to sea rise as it will face unbearable heat and its water sources will diminish. Already in Java, the island has only 12 percent of forest cover when it requires 25–30 percent of forest cover to maintain ecological equilibrium and water supplies. Bali, the jewel of Indonesian tourism, is facing severe water shortages and drought and one wonders in another five years if the island's water supplies will be able to supply its massive tourism population. The point of getting Indonesian leaders to see the detrimental effects of forest fires and resulting haze is not to concentrate on 'regional' impacts but to draw the national economic and social implications of Indonesia's forest fires in their own backyards. Indonesians must see that their forest fires and haze undermine the livelihoods of their own citizens and the sustainability of their development programs.

The human arena: Managing developmentalism

Domestically, the Indonesian government seems unable to manage its development programs. It wants to develop its large land areas productively but the benefits are not always going to its rural masses. The development of plantation agriculture in coffee, tea, oil palm, rubber and other commercial crops is heavily dictated by transnational companies and wealthy land owners. Riding on international market demand, Indonesian smallholders also get involved in growing lucrative crops. Sulawesi is a good example of small holdings' coffee production arising from global market demands. The Indonesian government abets such large scale developments for foreign revenue and to garner a global market share of each agricultural product. In palm oil, it is obvious that Indonesia wants to compete with Malaysia for the global market share. Hence, government officials have often turned a blind eye to forest clearance by fire to allow companies to develop palm oil plantations.

For domestic economic reasons, Indonesia has thus dragged its administrative feet in policing corporations and individuals clearing forest lands by fire. In the past, cronies of President Suharto like Bob Hassan were culprits of forest clearing fires but were beyond jurisdiction. Rampant corruption in the police, courts and security agencies allowed corporations and individuals to get away without getting caught or charged legally. The fact that the haze does not impact on Java and Jakarta makes for the political laxity to take haze control seriously. The demographic differences between Java and the Outer Islands also mean the political fallout in Kalimantan and Sumatra, the two most haze affected islands is less politically pressing. Since 65 percent of Indonesians live in Java–Madura, the Outer Island populations are often taken by the Jakarta power elites as politically expedient.

Reformasi (political reforms) and change in the post-Suharto period have paved the way for greater democracy and decentralization of power to provincial administrations and regencies. On the surface, this might look like a move towards an equalization of wealth, the closing of the gap between rich and poor. But in reality, it has created greater red tape in managing forest reserves. Corruption which was in the past under the Suharto regime centralized in the federal capital in Jakarta has now permeated to all levels down to the provincial and village levels — everyone wants a piece of the pie.

One can recall Zakaria's (2007) criticism of the US federal and democratic system. Specifically, the rise of US illiberal democracy has become the political order of the day in which every major constituent (Hispanics, Asians, Jews, African Americans, Whites, Muslims, corporations, rich, poor) in the US uses the democratic system to further its own interests at the expense of the national state. In Indonesia under reformasi, the federal governmental system has become divided into splintered provincial demands and autonomous political entities (Aceh) are no different. When natural resources within provinces are claimed by provincial governments as their wealth, the conflicts between the federal and provincial governments can become nasty, protracted and difficult to manage. Now the exploitation of natural resources goes through a larger number of hands of corruption from the federal to the community level because of administrative decentralization. On the surface it might look like operationalising democracy but in reality it has become a case of administrative involution — more administrators dividing the fixed pie amongst themselves.

The move under reformasi was a reversal of forest policies under President Suharto. The Suharto regime brought the traditional forest areas of the village *adat* (customary law) jurisdiction under the centralized Federal government jurisdiction. Reformasi rightly put forest areas back in the hands of the village *adat* (Lindayati, 2003). But villages are themselves under economic threat and hence rather than protecting their forest areas, they are exploited, utilized and sold commercially to upkeep poor families. Hence, forest fires in Indonesia are a product of democratic processes with forest fires emanating from many village *adat* controlled forest areas. At a macro-level, Indonesia and other developing countries would have to ask if Duara's (2015) proposition of achieving sustainable modernization is possible. Conceptually, it is an appealing proposition but operationally it is hard to implement and manage.

The international common-pool resources

Like most developing tropical countries, Indonesia is at the mercy of a global capitalistic system which creates an iron-clad financial scaffolding that is difficult to resist or change. It is difficult for poor developing countries to get out of the economic capitalistic trap in dealing with natural resources and raw materials. The economics of the trade-off between saving forest reserves and growing oil palm underscores the capitalistic logic. Ironically, a lot of the capitalistic logic on natural resources such as forests reserves is undergirded in Karl Marx's *Capital* thesis. He argued that the value of nature was based on 'exchange value' and it was wealth given as 'gratis' or a free gift. He viewed the earth as a 'reservoir' from where the 'use-values' are derived. Hence, it was Marx in his labor theory of value that argued that it was not nature but human labor which generated value (see Foster *et al.*, 2010: pp. 75–82).

This Marxian conceptual thinking of 'mother nature' has unfortunately undervalued the importance of Nature in the production of wealth. Though in economics today 'nature value' and wealth (generated by human labor) are conflated, there is still the underlying belief amongst certain economists, government leaders and businessmen to accept Nature as a given without any intrinsic value. Hence, it is logical for Indonesian politicians, bureaucrats and corporate titans to see 'forest reserves' as relatively worthless until there are converted into cash crop plantations of oil palm, coffee, tea or rubber which can be traded on the global market. The logic of burning down massive forests in Indonesia is not seen as a crime but a bonus of adding value to the country's wealth by converting 'worthless forest' into productive economic wealth.

On the other hand, the question of ascribing an economic value to forest often translates to a value of a current market value of land or timber which does not do justice to forest conservation. What is often missing is trying to ascertain the intrinsic value of forests reserves as part of a wider ecosystem with rich biodiversity. In many cases, virgin and pristine forests would have an undefinable value — because its future value to the wider national and global community is irreplaceable. While in the earlier forest fires and haze scenarios, blame was placed on shifting cultivators and plantation, in the 2015 forest fire blame shifted to paper and pulp companies as well. The chief target was the Indonesian conglomerate Sinar Mas, owners of Asia Paper and Pulp (APP), Golden Agri and Sinar Mas Land who operate the largest paper and pulp company in Southeast Asia.

Many environmentalists have thus written widely about the disjuncture between capitalism and nature conservation (see Foster, 2002; Speth, 2008; Foster *et al.*, 2010). In a developing country like Indonesia with extensive poverty, wide disparities of wealth, and a large population of 264 million to support, the economic propensity and political imperative to develop large tracts of forested lands into seemingly productive areas seem have to criticize and debunk. What adds insult to injury is when global conservationists chide governments in the developing countries to think Green, save the lungs of the world and conserve their forests for the global common good. The retort of Indonesians, Brazilians and Malaysians is why should poor countries be sacrificing their countries, tropical forests, economic development and standard of living so that the rest of the world can enjoy and prosper? As a response to former Malaysian Prime Minister Tun Dato' Seri Dr Mahathir Mohamad's pointed criticisms of the demands for tropical forest conservation, the developed countries had initiated REDD+ (Reducing Emissions from Deforestation in Developing Countries) which was meant to reduce CO_2 emissions from deforestation and forest degradation and foster conservation, and the sustainable management of forests. The idea was to pay for the conservation of tropical forests as 'lungs of the world'. Indonesia with support from Norway got a two-year moratorium from 2011 to restrict deforestation (Stern, 2015: pp. 72–73). Unfortunately the REDD+ program has been short-lived as developed countries face economic uncertainties due to a volatile global economic situation since the 1997 crisis and developing countries are unable to police illegal deforestation.

Yet there are other ways of dealing with reckless deforestation from the developed world's perspective which could help save tropical forest. In its 2015 annual report, Norway's Government Pension Fund Global (GPFG) for example has dropped 11 companies because of its connections with deforestation (Fogarty, 2016: A7). With 830 billion US dollars invested in 9,000 firms worldwide, corporations do take note of GPFG principles used in investments. In 2015, GPFG took corporate green responsibility seriously and divested in 73 companies related to deforestation and greenhouse gas emissions (Fogarty, 2016: A7). While developed countries have not been able to pressure Indonesia to save its tropical forests, two other developments as discussed earlier have given the Indonesian government food for thought about their forest conservation needs: global warming and the mass extinction of organisms.

The economic expediency of clearing lands cheaply through fire by both corporate companies and smallholders in Indonesia is not an easily solved problem. While several ASEAN countries (Malaysia, Singapore, Brunei, Thailand and the Philippines) have been the unsolicited victims of haze for decades due to Indonesia forest fires, the Indonesian government has shown less resolve to solve the problem. The Indonesian political leaders are less than sympathetic to its neighbors' criticisms. Indeed, Indonesian Vice-President Jusuf Kalla said that the countries in the region instead of grousing about the haze should thank Indonesia for giving them 11 months of good, clean air. The Indonesian political venom is particularly targeted at wealthy Singapore where the stark disparities of wealth between the two countries make the Singapore criticisms of its larger neighbor a politically sensitive issue.

Clearly Indonesian forest fires and its resulting haze are politically complex, difficult to manage and there is a lack of expertise to deal with the issue. Despite the economic fallout during the forest fires and haze within Indonesia, there is also ironically a silver lining. Every time *El Niño* strikes with a vengeance, oil palm production drops and palm oil prices surge the following year. The fall in production of oil palm in 1997–1998 and 2015–2016, both years of severe *El Niño* and resulting forest fires have seen a rise in cooking oil prices and vegetable oils used for chocolates, soaps and biofuels. Capitalism indeed thrives on scarcities.

Impacts on developmental trajectories

Indonesia, Malaysia and Myanmar are three Southeast Asian countries where forests reserves are abundant but these might all vanish in a couple of decades if their governments are not percipient and prudent in conserving their forests. The Indonesian government should look around the region and decide whether they want to destroy their national forest reserves like what has happened in Thailand, the Philippines and Cambodia. Vietnam has also lost its forest reserves but this is due to the long protracted Vietnam War where chemical warfare destroyed much of her forests.

From a global and regional standpoint, forests and oceans are the carbon sinks of the world as well as suppliers of oxygen. Given the dire warnings of CO_2 increases leading to global warming, Indonesia should take heed of saving her forest reserves for maintaining an ecological equilibrium in development. Eradicating forest is compounding already a massive trend in desiccation processes in the region. In early 2016, the high heat waves (40 degrees), prolonged drought and disruption of agricultural cycles had undermined livelihoods in mainland Southeast Asian countries as it did in Malaysia. Climate change is wreaking havoc in the region and the areas of mitigation are beyond the financial and economic abilities of countries in the less developed world. Every year when *El Niño* hits the region, governments are faced with insurmountable challenges. Poverty stricken villagers find ways to circumvent their personal misery but in the process create problems of a larger scale. Large companies exploit the situation for profits with short-sighted perspectives.

Unfortunately in the economic development of countries, it would seem that land evaluated as 'property' is more important than saving 'useless' forested areas. Growing population, disparities of wealth and finite natural resources are putting stains on the developmental goals and programs of a large country like Indonesia. Despite having 40 percent of the region's gross national product (GNP), the country also has to feed 40 percent of the region's population.

Reflections

The Indonesian fires and haze provide a powerful demonstration of how nature abetted by human activities and economic greed can lead to major landscape changes, human casualties, and economic costs. Is this an ephemeral blip or a major environmental trend in Southeast Asia? It might be too late to counter this problem if the *El Niño* droughts, desiccation and fires are a climatic trend and not just weather changes. The people in Southeast Asia are going to be victims of a larger global climatic issue. It would seem that in times of personal desperation and economic opportunity when environmental challenges unfold people think of their survival first or selfish economic motivations and less about the wider ramifications of their actions on the community and ecology.

Seen within a longer time scale and wider spatial perspective, the prolonged droughts, forest fires and the haze which are products over the last three decades are best seen as symptoms of not weather but climate change — something which some academics, politicians and scientists are hesitant to accept. Given the quest for national sustainability and personal survival, one can expect the unfolding problems and challenges which climate change brings will only accelerate more drastic, draconian and desperate measures by governments, corporations and families. The fall-out of forest fires and resulting haze is only the tip of the iceberg of the deep metabolic fracture in human–nature relations (Foster *et al.*, 2010) and the fact we are drawing closer to the "tipping point" (Flannery, 2009) in the unfolding global warming scenario. In his book *Atmosphere of Hope*, Tim Flannery (2015: p. 28) argues that we are in the "age of the mega-fire" where annually 300,000 die by inhaling smoke from forest fires.

The crux of this environmental crisis ironically is no more a product of Nature's unpredictable wrath per se but also human motivations. With the weakening of spiritual human–nature bonds which helped to bind indigenous peoples closer to Nature, the reverence and respect for nature are splintered in a world dictated by science and technology. Science has decoded the environment into elements, molecules and atoms and this has undermined the holistic understanding of the importance of ecosystem logic. For example, one criticism of Stern's economic understanding of climate change is that he uses rather simplistic laboratory notions of the greenhouse effect based on 'glass based plant greenhouses' whereas in the global environment CO_2 has a more complex relationship with atmosphere, water vapour, vegetation and the oceans.

We cannot decode and deconstruct Nature as though it is a machine or robot. What we require is not Newtonian science in understanding Nature but the science of Goethe which provides a more inclusive, phenomenological and holistic appreciation of Nature. The issue of forest fires and the haze in the region cannot be solved by applying Newtonian science through quantitative principles alone. What is required is the science of Goethe, which depends on a holistic 'organic' perspective of the "multiplicity in unity" — allowing for differences and uniqueness (Bortoft, 1996: 92–99). This means the communities living in Kalimantan and Sumatra must be engaged in the development of their cultural landscapes in a more engaged and environmentally sustainable manner as opposed to scientifically prescribed plantation agriculture which is product orientated and profit motivated and cares little about ecosystem sustainability. Such indigenous communities have always believed in Wilson's (1984) Biophilic relationship with nature. From a sustainable perspective, indigenous forms of land tenure and agriculture maintain a 'relational' behavior to the environment (Peterson, 2001) which Diamond (2012) demonstrated in the Pacific Islands and Papua New Guinea has been sustained for hundreds of years. When corporations and plantations owners deal with forest, they unfortunately isolate and discriminate Nature in economic terms as monetary value and profits and forget the broader ecological and organic relationships of forest to the rest of nature and human beings.

As we enter the new geological phase of the Anthropocene, one cannot assume that human induced environmental disasters are place and time specific — our ecosystems are changing with human interventions and one wonders what the future global ecosystem would be like given the expanding human population coupled with its destructive technology and the quenching hunger for natural resources. Indonesia is a model reflection of a developing country with a vast human population that is poor and with wide wealth disparities seeking to improve the quality of living. Without proper environmental management principles in the context of eco-development, forest fires are likely to remain a convenient excuse in land expansion, agricultural change and economic development. In short, what this chapter is arguing is that forest fires which are raging throughout the developed world in Australia, Europe and America (California, Alaska) as a result of global warming is not unique to Indonesia and the Southeast Asian region. As a global community, one has to accept that forest fires and transboundary haze have become yet another negative outcome of climate change which many countries have difficulty managing much less solving and eliminating. This puts the onus on governments to put priority in tackling climate change. Of all countries in the Southeast Asian region, Indonesia is the highest emitter of CO_2. To mitigate high greenhouse gas emissions, the Indonesian authorities need to manage carefully their land clearance, forest fires, agricultural change, urban development and transport systems.

References

https://en.wikipedia.org/wiki/2011_Thailand_floods (Accessed: June 1, 2017).

https://www.forestcarbonpartnership.org/what-redd (Accessed: June 1, 2017).

Asian Development Bank. (2001). *Asian Environment Outlook 2001*, Asian Development Bank: Manila.

Bender, M. L. (2013). *Paleoclimate*, Princeton University Press: Princeton.

Benjamin, G. (2002). On being tribal in the Malay World, in Benjamin, G. and Chou, C. (eds.), *Tribal Communities in the Malay World*, ISEAS: Singapore, pp. 7–76.

Bortoft, H. (1996). *The Wholeness of Nature*, Lindisfarne Books: Great Barrington, US.

Diamond, J. (2012). *The World until Yesterday*, Allen Lane: London.

Duara, P. (2015). *The Crisis of Global Modernity*, Cambridge University Press: Cambridge.

Flannery, T. (2006). *The Weather Makers*, Grove Press: New York.

——— (2009). *Now or Never*, Atlantic Monthly Press: New York.

Fogarty, D. (2016, March 28). Norway fund drops 11 firms over deforestation. *The Straits Times*, p. A7.

Foster, J. B. (2002). *Ecology against Capitalism*, Monthly Review Press: New York.

Foster, J. B., Clark, B. and York, R. (2010). *The Ecological Rift*, Monthly Review Press: New York.

Geertz, C. (1963). *Agricultural Involution: The Processes of Ecological Change in Indonesia*, University of California Press: Berkeley.

Giddens, A. (2009). *The Politics of Climate Change*, Polity Press: Cambridge, UK.

Glacken, C. (1967). *Traces of the Rhodian Shore*, University of California Press: Berkeley.

Kolbert, E. (2014). *The Sixth Extinction*, Bloomsbury: London.

Lieberman, V. (2003). *Strange Parallels: Southeast Asia in Global Context, c. 800–1830,* Vol. 1, Cambridge University Press: Cambridge.

———— (2009). *Strange Parallels: Southeast Asia in Global Context, c. 800–1830,* Vol. 2, Cambridge University Press: New York.

Lindayati, R. (2003). Shaping local forest tenure in national politics, in Dolsak, N. and Ostrom, E. (eds.), *The Commons in the New Millennium*, The MIT Press: Cambridge, Massachusetts, pp. 221–264.

Linden, E. (2007). *The Winds of Change: Climate, Weather, and the Destruction of Civilizations*, Simon and Schuster: New York.

Lovelock, J. (2007). *The Revenge of Gaia*. Penguin Books: London.

Ohara, T., Akimoto, H., Kurokawa, J., Horii, N., Yamaji, K., Yan, X. and Hayasaka, T. (2007). An Asian emission inventory of anthropogenetic emission sources for the period 1980–2020, *Atmospheric Chemistry and Physics*, 7: pp. 4419–4444.

Parker, G. (2013). *Global Crisis*, Yale University Press: New Haven.

Peterson, A. L. (2001). *Being Human*, University of California Press: Berkeley.

Rosen, W. (2014). *The Third Horseman*, Viking: New York.

Sim, M. (2015, September 29). More international cooperation needed to tackle haze: Vivian. *The Straits Times*: p. A8.

Speth, J. G. (2008). *The Bridge at the Edge of the World*, Yale University Press: New Haven.

Stern, N. (2015). *Why Are We Waiting?* The MIT Press: Cambridge, Massachusetts.

Tan, Y. S. (2016, March 21). Chiang Mai's headache: Corn-fed smoke haze. *The Straits Times*, p. A11.

Whyte, I. (2008). *World without End? Environmental Disaster and the Collapse of Empires*, I. B. Tauris: London.

Wilson, E. O. (1984). *Biophilia*, Harvard University Press: Cambridge, Massachusetts.

Zakaria, F. (2007). *The Future of Freedom*, W.W. Norton & Company: New York.

18 Can Responsible Corporate Behavior Clean up the Haze?

Asif Iqbal Siddiqui

Curtin University Sustainability Policy (CUSP) Institute, Western Australia
ai.siddiqui@curtin.edu.au

Background

Transboundary haze originating from Indonesian forest fires has been a longstanding issue in Singapore and Southeast Asia as a region. Public and media attention as early as the 1970s has been drawn on the effect on transboundary pollution resulting from the Indonesian forest fires, although policy makers only started paying closer attention to the issues since the mid-1990s leading to a number of [public sector/private sector] initiatives (Quah, 2016). Nonetheless after two decades, Singapore and the region have continued to be stifled by the smog. Indeed, the problem has increased beyond proportion well into hazardous levels in recent years. In 2013, pollution standard index (PSI) in Singapore went above 400, which to date is recorded as the historic highest level and in 2015–2016 the city-state experienced the most lasting episode of haze. Therefore, it is timely to revisit the existing policy regime which accentuates the traditional role of government regulations and its efficacy.

This chapter argues that there are at least two key limitations to the traditional approach. First, since the source of haze is forest fires in Indonesia, the onus is on the Indonesian government to manage the fires in its backyard, which in turn is handicapped by decentralized governance, resource limitations and administrative corruptions. Second, the regional governments of the affected countries like Singapore or Malaysia can only diplomatically pursue Jakarta through non-binding international agreements. At best, the affected regional governments can assist Jakarta with financial, technological and logistical issues. Hence, here I would like to take the focus away from the central role of political governance to the economic role of corporate governance especially with reference to large corporations involved in forestland clearing with fire in Indonesia.

More specifically, as the focus of this chapter suggests, it argues that responsible corporate behavior could be a potent way to address the issue. The approach assumes the involvement of three interdependent stakeholders in the strategic game, namely corporates, community, and government. While the role of governments is predominantly confined within the national boundary and inter-governmental diplomatic activities, the role of community and corporations is more pervasive and versatile. Large-scale forest fires causing haze can be linked with the large-scale industrial activities, therefore engaging industries would be the most potent solution as much as they are the main source of the problem. Meanwhile, governments can target industries through regulatory measures and sanctions, and the citizens who legitimise government's regulatory roles also represent the consumers and investors of the industries in question. Hence, the role of the community can be very influential.

A wicked problem

It is not surprising that given the intensity and complexity of the problem, that an effective or indeed, any solution remains fairly elusive. The indigenous communities, especially in the large islands of Riau, Sumatra and Borneo have been using slash-and-burn methods for centuries to clear forest lands. However, the scale of fire and thereby the intensity of the smog started to reach an alarming level with the rise of land clearing for commercial usage especially since the 1950s. The scale of fire, reflected in the volume of smoke, started to travel beyond the Indonesian archipelago. Thus, its scale and complexity have grown beyond the scope of Indonesian community, industries and government.

Scale and cost of fire and haze

According to a recent World Bank report, 2.6 million hectare of forestland was cleared by fire in 2015, alone which included the peat swamp forests with thick layers of acidic peat and high carbon.[1] In addition to the destruction of forest ecosystem and biodiversity, the fire generates enormous volume of haze often traveling across the region. In 2015, September–October volume of CO_2 emission from the forest fires exceeded the daily average of the US, which is 15.95 million tons according to World Resources Institute.[2] The smoke can persist in the air for months causing a range of problems from discomfort and low visibility to respiratory illnesses. In Indonesia alone, it affected 28 million people that include 136,000 suffering from acute respiratory infections.

The cost of forest fires and haze has been measured in dollars for the first time during the 1997 episode, which was estimated at 25.4 billion US dollars. Indonesia had to share 93.8 percent of the cost while Malaysia and Singapore had to share 5.1 percent and 1.1 percent respectively (Siddiqui and Quah, 2004). In 2015, according to Indonesia's National Disaster Management Agency (BNPB) the government spent 385 billion rupiah on tackling fires. Jakarta also lost 12.5 billion rupiah in timber loyalties leading to a total cost of 475 trillion rupiah.[3] In another estimate by The World Bank, Indonesia incurred total damage worth 16 billion US dollars due to the fires and haze.

Root cause analysis

The root cause analysis of haze reveals that there are multiple layers to the problem involving multiple layers of stakeholders. Land clearing by fires is a deliberate economic decision by farmers, smallholders and industries as it is easier, faster and more cost effective. It is done especially during the dry seasons which increases the risk of the fire going out of control. The regional weather pattern such as La Nina and El Niño Southern Oscillation then prolongs the fires as observed in 1997 and 2015. As local communities have been using fire to clear land for centuries, there is a common tendency to blame them for haze, although the large-scale fires

---

[1] http://www.straitstimes.com/asia/se-asia/haze-fires-cost-indonesia-22b-twice-tsunami-bill-world-bank

[2] http://www.channelnewsasia.com/news/asiapacific/carbon-from-indonesia/2207404.html

[3] http://www.straitstimes.com/asia/47b-indonesia-counts-costs-of-haze

that are the main source of transboundary haze across the region are essentially caused by large industries. Hence, commercial usage of the land by the large corporations must be brought under the spotlight.

The industries heavily involved in deforestation include palm oil, pulp and paper, and timber processing. Indonesia is one of the largest palm oil producers in the world. The growth of palm oil industry has been phenomenal as it is used in a wide range of products from biscuits to shampoo. Using the national level data on land use between 1975 and 2005 (Wike *et al.*, 2011), they have found that 40 million hectare forest cover has been lost in Indonesia representing around 30 percent reduction of forest cover mainly due to growing demand for palm oil. However, the value chain of these industries is not limited to Indonesia. They have consumers and investors around the region and the world. A few producers are listed as public companies. Thus, there is an enormous potential to effectively engage these corporations through engaging investors, the distribution channels, consumers along with imposing regulations.

Public diplomacy and policy

Forest fires in Indonesia have been a regular annual event. Transboundary haze usually receives attention from media when it is intense and visible. Although, regional haze episodes have appeared in media on and off since 1970s, significant policy responses came only in 1995 when Association of South East Asian Nations (ASEAN) Cooperation Plan on Transboundary Haze Pollution was adopted, leading to the Regional Haze Action Plan in 1997. In line with the culture of centralization of power, there is high dependence on the government regulations in the public space in ASEAN. That often explains why there is a lack of market-based instrument addressing haze or other environmental issues.

Since forest fires in Indonesia causes the transboundary haze, by default it becomes Indonesia's obligation to regulate the activities and the parties causing and contributing to the haze. Indonesia also bears the burden of the lion share of the cost from fire and haze. Thus, it also makes economic sense for Jakarta to take action. The government however requires necessary administrative support and adequate finance to monitor the activities and enforce regulations. It would be toothless otherwise. In order to address the resource constraints of Jakarta, affected neighbors have also been providing satellite and meteorological data to identify the fire hotspots as well as providing aircraft and logistics for cloud seeding to control fire among other sources of assistance especially over the last two decades.

The necessary administrative support is tied with the system of governance, legal norms, and land entitlements necessary for monitoring and managing the hotspots. The Indonesian government is highly decentralized with regional and local governments often holding the authority to provide land tenure and monitor land usage. It therefore becomes difficult for Jakarta to get timely and adequate information about what is happening in remote places in Riau, Kalimantan or Sumatra. Jakarta has enacted regulations like the Environmental Management Law (No. 23, 1997) to protect the forests with high penalties. It is also difficult to enforce the laws due to lack of land tenure information as well as corruption and inefficiency of the local government system. Thus, despite the availability of the data on hotspots, which can quite accurately identify the origin of fire, it has been challenging to identify and sanction the perpetrators initiating the fire.

It is worth noting that there might be a vested interest for Indonesia to protect local industries. From an objective cost-benefit analysis in support of industries versus the cost to Indonesia in mitigating the effects of forest fires, according to the World Bank report if each of 2.6 million hectare forest land cleared in 2015 is converted for palm oil planation its value would only be 8 billion US dollars which will go to private hands whereas the cost of the fire to Indonesian economy is 16 billion US dollars.[4] In other words, there will be a net welfare (deadweight) loss of 8 billion US dollars. The industrial benefit does not appear to justify the cost to Indonesia and therefore would support the proposition of government intervention through regulatory action.

Governance of corporations

The regulatory role of political governance is essential in the socio-political environment of ASEAN despite its limitations, although it must

---

[4]http://www.straitstimes.com/asia/se-asia/haze-fires-cost-indonesia-22b-twice-tsunami-bill-world-bank

be complemented with the necessary corporate governance. Regulations directly targeting the corporations involved in forest fires, along with the active role of investors and consumers, are necessary to address a complex issue like transboundary haze. The large corporations involved in land clearing often have significant clout not only with central governments but also with local governments and communities. Therefore, it has been difficult to get local evidence on their involvement in illegitimate forest fires. Only in recent years have the regulators started to scrutinize their activities closely.

In 2015, Jakarta finally initiated a series of measures to bring the guilty corporates to justice. The authorities arrested executives from seven companies allegedly responsible for causing illegal fire under the law against environmental mismanagement.[5] If convicted, they could receive jail sentences of up to 10 years and fines of 10 billion rupiah. Police has reported that 20 more companies and 140 individuals have also been under investigation.[6] However, no one has been held responsible yet, although the arrest of the senior executives depicts significant commitment from the government to seriously redress misdoings. The court in Sumatra has dismissed the civil suit against Bumi Mekar Hijau for 7.8 trillion rupiah, although the government still intends to appeal against the decision in the Supreme Court.[7]

It is worth noting that these large corporations often operate regionally and their value chain is subject to multiple jurisdictions. Hence, the regional governments can work together to regulate them without having to worry about the non-binding nature of international agreements and treaties. In 2015, for example Singapore commenced a legal process against five listed companies revealing their identities for the first time.[8] Singapore's National Environment Agency (NEA) has begun gathering evidence against these

---

[5] http://www.channelnewsasia.com/news/singapore/indonesia-arrests-seven/2132714.html
[6] http://www.eco-business.com/news/indonesia-gets-tough-on-companies-responsible-for-haze/
[7] http://www.straitstimes.com/asia/se-asia/indonesia-to-appeal-courts-rejection-of-lawsuit-against-firm-over-haze-causing-fires
[8] http://www.straitstimes.com/singapore/environment/spore-clamps-down-on-five-firms-over-haze

entities through satellite images and meteorological data. For example, Asia Pulp and Paper has been asked to provide information on Indonesian value chain which includes four Indonesian companies — Rimba Hutani Mas, Sebangun Bumi Andalas Wood Industries, Bumi Sriwijaya and Wachyuni. Indonesia meanwhile initiated legal process against them too. Under the provisions of Transboundary Haze Pollution Act 2014, guilty companies could face fines of up to 100,000 Singapore dollars a day (capped at 2 million Singapore dollars. Meanwhile, the Malaysian court has held an executive of PT ADEI Plantation, a unit of Kuala Lumpur Kepong Berhad, responsible for the fire in Riau and sentenced him to one year in prison or an option to pay 2 billion rupiah in compensation.[9]

Consumer and investor activism

In order to engage corporates in responsible behavior, consumer and investor activism has been increasing around the world (Christofi *et al.*, 2012). The collective investment power of the socially responsible investors especially since 1970s has made significant impacts forcing companies to rethink their business risk associated with environmental, social and governance (ESG) issues which led to the notion of socially responsible investments (SRIs) since 1980s (Lashgari and Gant, 1989; Sanyal and Neves, 1991). However, the movement has received little to no momentum so far in Asia. Here it has usually been the environmental interest groups and non-government organizations (NGOs) acting at the forefront of pressuring industries to comply with the environmental standards,[10] although as rational entities, the industries are more likely to take consumers and investors more seriously.

It is nevertheless interesting that civil movement against irresponsible corporate behavior is gaining some momentum here. In 2007, for example a group called Haze Elimination Action Team (HEAT) was formed. The HEAT has been advocating for the boycotting of products produced by

---

[9]http://www.thejakartapost.com/news/2014/09/11/malaysian-firm-fined-executives-get-prison-role-forest-fires.html

[10]http://www.channelnewsasia.com/news/singapore/indonesia-arrests-seven/2132714.html

companies responsible for contributing to the haze. It has also been organizing funds to file class action lawsuits against perpetrator companies.[11] The Consumers Association of Singapore (CASE) has become a member of Consumer International (CI), which has also been advocating for consumers to boycott wood, paper or pulp product from the companies allegedly involved in forest fires in Indonesia.[12] The largest retail chain in Singapore NTUC Fair Price and 16 other firms signed a Haze-Free Declaration that their products will not have any links with firms responsible for causing haze, in response to a call from the Singapore Environment Council (SEC). The products are related to pulp and paper. Tan *et al.* (2009) suggest palm oil industry has made significant contribution in deforestation, peatland destruction and degradation and depletion of biodiversity leading to some NGOs lobbying for the boycott of palm oil products. However, SEC is yet to work on to address the products related to palm oil.[13]

Unfortunately, a significant number of consumers and retail investors are not aware of socio-environmental consequences of their consumption and investments. In a street poll conducted by *The Straits Times* in Singapore, only 50 percent of the respondents expressed their desire to boycott haze-related products as many do not recognize that it would make a difference.[14] It is also believed that Singapore is too small as an economy to make an impact although it has a substantial clout in terms of purchasing power.

Corporate responsibilities

Responsibility of corporates to the society has been recognized and emphasized since 1950s, although corporate social responsibility (CSR) has been popularized in 1990s by advocates of SRI and sustainable development

---

[11] http://www.theonlinecitizen.com/2015/09/sg-govt-gets-tough-with-errant-companies-responsible-for-haze/

[12] http://www.straitstimes.com/singapore/environment/consumers-urged-to-boycott-products-from-companies-linked-to-haze

[13] http://www.straitstimes.com/singapore/environment/more-firms-on-board-for-haze-free-declaration?login=true

[14] http://www.straitstimes.com/singapore/environment/many-will-boycott-products-from-haze-causing-firms-poll

(Carroll and Shabana, 2010). Consequently, many corporates started socially responsible activities, corporate social responsibility (CSR) reporting and publishing Supply Chain Responsibility (SCR) reports annually. In 2007, Indonesia has made CSR mandatory for the listed companies (Law No. 40). The CSR is generally understood as being evidenced by donations and/or community development expenses while the mandatory provision of CSR violates its voluntary spirit (Waagstein, 2010). However, companies often use CSR in reducing conflict with the community while continuing business as usual with little regards to sustainability. For example, the palm oil companies use CSR activates like education, health care and local infrastructure developments to reduce conflict with smallholders and community (Gunawan *et al.*, 2009; Sugino *et al.*, 2015).

Similarly, corporations tend to use CSR reports as an opportunity to enhance their corporate image. The culture of CSR reporting thus has two major limitations. First, as mentioned it has been primarily used as a tool only for enhancing corporate image and greenwashing the business. Second, CSR being viewed as a charity initiative unrelated to sustainability of the core business and which therefore does not appeal to investors. In developed countries traditional CSR reporting has often been replaced by more comprehensive integrated reporting (IR) aiming at reporting companies' sustainability performance and exposure to ESG risks. Global Reporting Initiative (GRI) framework for reporting ESG performance is widely used around the world for producing both traditional CSR reports and IR while the countries usually publishing IR have stronger market orientation, disperse ownership structure, higher trade union densities, and higher level of socio-environmental development (Jensen and Berg, 2012). In Indonesia, Malaysia and Singapore GRI framework is used usually for CSR reporting with little attention to overall sustainability aspects and ESG risk exposure which the regulators and investors require to reconsider. Thus, reliance on CSR or CSR reporting can indeed be a distraction for the regulators, consumers and investors.

Conclusion

Transboundary haze is essentially a problem which has a regional impact and which is rooted primarily in large-scale land clearing with fire for commercial usage. The traditional policy regime with command and control mechanism is essential in Asian culture of governance, although it has several limitations. In a large country like Indonesia with highly decentralized governance and prevalent corruption, the central government has to face multiple layers of agency problems and moral hazards in implementing policies. Thus, Jakarta might find it difficult to satisfy the regional neighbors' demand for effective forest fire management. It is however convincing that large corporates can be involved in large-scale fires causing transboundary haze. Furthermore, as their value chains are often extended in the regional market active consumers and investors can play a vital role in assuring that the corporates behave responsibly. The regulators and market participators therefore have to take necessary steps to remind corporates of their real exposure to ESG risk. Nevertheless, it requires development of a culture in the region where consumers and investors will be more vigilant about the socio-environmental consequences of their consumption and investment on their health and wellbeing.

References

Carroll, A. B. and Shabana, K. M. (2010). The business case for corporate social responsibility: A review of concepts, research and practice, *International Journal of Management Reviews*, 12(1): 85–105.

Christofi, A., Christofi, P. and Sisaye, S. (2012). Corporate sustainability: Historical development and reporting practices, *Management Research Review*, 35(2): 157–172.

Gunawan, J., Djajadikerta, H. and Smith, M. (2009). An examination of corporate social disclosures in the annual reports of Indonesian listed companie, *Centre for Accounting, Governance and Sustainability*, 15(1): 14–36.

Jensen, J. C. and Berg, N. (2012). Determinants of traditional sustainability reporting versus integrated reporting: An institutionalist approach, *Business Strategy and the Environment*, 21(5): 299–316.

Lashgari, M. K. and Gant, D. R. (1989). Social investing: The Sullivan principles, *Review of Social Economy*, 47(1): 74–83.

Quah, E. (2016). Economics of transboundary haze pollution in Southeast Asia: Is there a solution? 12th International Conference of Western Economic Association International, Singapore.

Sanyal, R. N. and Neves, J. S. (1991). The Valdez principles: Implications for corporate social responsibility, *Journal of Business Ethics*, 10(12): 883–890.

Siddiqui, A. I., and Quah, E. (2004). Modelling transboundary air pollution in Southeast Asia: Policy regime and the role of stakeholders, *Environment and Planning A*, 36(8): 1411–1425.

Sugino, T., Mayrowani, H. and Kobayashi, H. (2015). Determinants for CSR in developing countries: The case of Indonesian palm oil companies, *The Japanese Journal of Rural Economics*, 17(0): 18–34.

Tan, K. T., Lee, K. T., Mohamed, A. R. and Bhatia, S. (2009). Palm oil: Addressing issues and towards sustainable development, *Renewable and Sustainable Energy Reviews*, 13(2): 420–427.

Waagstein, P. R. (2011). The mandatory corporate social responsibility in Indonesia: Problems and implications, *Journal of Business Ethics*, 98(3), 455–466.

Wicke, B., Sikkema, R., Dornburg, V. and Faaij, A. (2011). Exploring land use changes and the role of palm oil production in Indonesia and Malaysia, *Land Use Policy*, 28(1): 193–206.

19 All the World's a Game: And All the Men and Women Merely Players. Or Not.

Christabelle Soh

Editor, Economics & Society
www.econsandsociety.com

This chapter attempts to show how game theory, specifically cooperative games, can be applied to solve the Southeast Asian (SEA) transboundary haze pollution problem and the limitations of this approach. It concludes with preliminary ideas of how some of these limitations may be overcome.

Why a game theoretic approach?

The issue of pollution is not new to economists. Almost a century ago,[1] the economist, Pigou, wrote *The Economics of Welfare*[2] which introduced the concept of using taxes to correct external costs caused by pollution. The intuition was that a rational industrialist would not care about the damage to others caused by the pollutants from his factory unless he is made to pay for it. A tax on the producer would force him to account for such costs since he now has to pay for them and as such, reduce his production. Consequently, the amount of pollution generated would decrease.

In short, in the presence of pollution, one simply needs to estimate the cost of the pollution and levy a tax on the producer that is equal to this cost. Thus the Pigovian tax was conceived. Problem solved.

But not quite.

---

[1] 97 years, to be exact.
[2] Pigou, A. C. (1920). *The Economics of Welfare*, Mcmillan and Co: London.

The idiomatic fly in this ointment is this — to levy a tax, one needs to have jurisdiction over the producer. This is clearly not the case for pollution that crosses national boundaries. The Singapore government cannot levy a tax on an Indonesian firm, for example.

As such, barring the creation of a supranational body with jurisdiction over the whole Association of Southeast Asian Nations (ASEAN), the standard economic tool of imposing a Pigovian tax to correct negative externalities cannot be applied to the SEA transboundary haze problem. This leads us to look for solutions elsewhere.

And, that elsewhere is the field of game theory. To see how game theory is relevant, we need to understand what makes up a game[3]:

(1)   A game must have two or more players;
(2)   Each player must have strategies/choices;
(3)   The combination of strategies/choices must result in various outcomes/payoffs to each player.

It is clear that the SEA transboundary haze issue does fit the definition of a game: two or more players (countries) are involved; countries have strategies (e.g., Indonesia can decide whether and how much to clamp down on polluting firms, and Singapore and Malaysia can determine whether and how much to aid Indonesia); and the combination of strategies will result in different payoffs (i.e., different levels of haze pollution and wealth depending on the actions of each country).

Since the SEA transboundary haze issue does have the elements of a game, it is natural to look to game theory for insights into a possible solution.

All the world's a game — Or at least, all of SEA

It should be noted that with the exception of zero-sum games,[4] games are not actually classified into competitive and cooperative games. Rather, for

---

[3]For a formal definition of a game, please look up an introductory text on game theory. This chapter is not meant to be a technical discussion.

[4]Games in which the outcomes are such that the gain by one player necessarily means an equivalent loss by another.

any given game, there are two methods of analysis — competitive game analysis focuses on the strategies that each player would undertake to maximize his own payoff and the subsequent equilibrium that would be reached when all players do so while cooperative game analysis focuses on how incentives can be designed such that players will adopt the strategies to reach the outcome with the maximum total payoff.[5]

In the case of the SEA transboundary haze problem, there is definitely room to apply cooperative games analysis as the sum of benefits from the reduction in haze (e.g., the avoided damage from haze such as health costs) definitely outweighs the costs of haze abatement (e.g., the loss of income for certain firms) to the region. However, although the regional benefits outweigh the regional costs, the benefits and costs are not evenly spread out. In the absence of cooperation, the costs of haze abatement would likely be fully incurred by Indonesia while the benefits would be enjoyed by all. As such, a rational Indonesia would only engage in haze abatement to the degree that it benefits directly from it. Unsurprisingly, such efforts would be significantly less than it would be if Indonesia takes the benefits of haze abatement to other countries into account.

To incentivize Indonesia to switch to a strategy of higher haze abatement, a solution offered by cooperative game theory is that of side payments. If the damages caused by the haze to Singapore are 1 billion US dollars, Singapore could offer up to 1 billion US dollars to Indonesia as a 'reward' for haze abatement. If the earlier premise that the sum of benefits from the reduction in haze outweighs the costs of haze abatement is accurate, then the sum of available side payments from each country should be sufficient for Indonesia to reduce haze to the optimal level.

In fact, cooperative game theory also offers a framework to operationalize the use of side payments. For cooperation (including side payments) to be stable, three conditions must be met:

First, each country should be better off after the side payment as compared to before. In other words, the sum that Singapore and Malaysia must individually pay has to be less than the damages from the haze that they

---

[5]This also explains why cooperative games are irrelevant to zero-sum games. In zero-sum games, every combination of strategies has the same sum of payoffs as a gain (higher payoff) by one player is an equivalent loss (lower payoff) by another.

would each otherwise incur. In the same vein, the sum of payments that Indonesia receives must exceed the costs of haze abatement that is incurred.

Second, no group of countries can be made better off not cooperating than cooperating. This would be the case if there were multiple polluters and a separate agreement between some of the countries and some of the polluters would make that subgroup better off. Intuitively, this makes sense. If an alternative exists such that some of the countries would be made better off, then there would be little incentive for them to cooperate with the larger set of countries.

Third, the outcome should be Pareto optimal. This means that the arrangement of side payments must be such that there is no other arrangement that would improve the payoffs for any country without worsening the payoffs for others.

Working out a system of payments that satisfies the three conditions from scratch is a trying task. Fortunately, in a paper published in 1995,[6] economists Chander and Tulkens developed a means to determine how the costs of pollution abatement could be shared (the share of the costs provided by a non-polluting country can be thought of as a side payment to the polluter that has to take steps to reduce the pollution) such that all the three conditions are met. The Chander–Tulkens rule is a simple one. The conditions would be met if the cost of the haze abatement is shared in proportion to the damages of haze that would be incurred otherwise. While this chapter will not go into the mathematics of it, the intuition is clear — those with the most to lose should pay the most.

In a more recent article in *The Straits Times*,[7] Chander and Quah made a rough estimate of the costs that Singapore, Malaysia, and Indonesia should bear if the Chander–Tulkens rule is applied. By their estimates, the annual cost that each should bear are 13.2 million US dollars, 61.2 million US dollars, and 1.125 billion US dollars respectively. In terms of side payment, this means that Singapore and Malaysia would make side payments to Indonesia of 13.2 million US dollars and 61.2 million US dollars

---

[6] Chander, P. and Tulkens, H. (1995). A core-theoretic solution for the design of cooperative agreements on the transfrontier pollution, *International Tax and Public Finance*, 2(2): 279–293.

[7] Chander, P. and Quah, E. (2014). Tackling haze with cost-sharing. *The Straits Times*. Singapore Press Holdings.

annually since the costs of haze abatement would most likely be incurred by Indonesia.

In sum, the application of the Chander–Tulkens rule from the field of cooperative game theory suggests that a reasonable solution to the problem of the SEA transboundary haze problem is in sight.

But it's not all fun and games

If the world was run by economists, it would seem that an end to the transboundary haze is in sight. One simply has to estimate the damages from the haze, the costs of abatement, and apply the Chander–Tulkens cost-sharing rule.

But, and fortunately so, the world is not run by economists. And even if it were, there is still the issue of free ridership.[8] If Singapore is part of the cost sharing agreement and pays Indonesia to reduce haze pollution, then Malaysia also gets to enjoy the reduction in haze and at no expense. As such, there is little incentive for Malaysia to share in the cost and make side payments to Indonesia.

Free-riding is a non-issue for the Chander–Tulkens rule to work only if we assume that the moment one country declines to cooperate, all countries will also decline to cooperate and so the choice faced by any country is either to cooperate or incur the damages from nobody doing anything about haze abatement. In other words, there is no possibility of free-riding in this scenario as the only two possible outcomes are everybody cooperating and achieving haze abatement, or nobody cooperating and zero haze abatement. The option of free-riding, which requires one country to opt out and free-ride upon the efforts of the remaining countries' cooperative efforts to reduce haze, is assumed away. However, that may not be a reasonable assumption. It is possible that a country may choose to not cooperate and free-ride, and bet that the remaining countries would go ahead anyway. Such calculations could lead to the dissolution of a cooperative agreement.

---

[8] Ironically, economists seem to be more prone to free-riding than non-economists. See Marwell, G. and Ames, R. E. (1981), Economists free ride, does anyone else? Experiments on the provision of public goods, *Journal of Public Economics*, 15(3): 295–310.

Additionally, there are non-economic concerns as well.

The economic approach to the problem hinges on adjusting incentives to encourage cooperation and assumes that incentives are all that people respond to. However, and again fortunately so, homo sapiens also respond to perceptions of equity and fairness. There have been enough studies on ultimatum games[9] to establish the truth that a reductionist approach that simplifies decision making to a simple study of incentives is incomplete. Homo sapiens are willing to forego some measure of economic benefit in exchange for fairness. And, outside of economic laboratories, we have seen how issues of perceived fairness can cause logjams in environmental negotiations: one of the sticking points in climate change negotiations was that of fairness — developing countries wanted developed countries to take historical responsibility for the existing carbon emissions in the atmosphere even though it was clear that it was (and is) cheaper to reduce carbon emissions from the developing countries than it is for the developed. While not a perfect parallel, the broader point illustrated still holds — perceived fairness in the negotiations could pose a problem to the SEA transboundary haze problem as any economically sound solution must involve some measure of side payment to Indonesia, and this will definitely not sit well with segments of Singapore's and Malaysia's population whose ideas of fairness do not include paying the polluter to stop polluting.

There is also the issue of feasibility. The inter-country analysis assumes that each country has the ability to monitor and regulate the ongoings within its geographical boundaries. This may be possible in small states like Singapore but may prove to be difficult for large countries like Indonesia. For Indonesia to have control over haze abatement, it needs to be able to detect fires, ascertain the causes, and inflict the appropriate punishment on the guilty parties. There is much work to do here. For one, to identify a guilty party, there needs to be a clear legal framework on who bears the liability for a forest fire. Should a natural fire break out,[10] is the landowner obliged to put it out? Or is the landowner

---

[9]Bearden, J. N. (2001). Ultimatum bargaining experiments: The state of the art. Available at SSRN: http://ssrn.com/abstract=626183 or http://dx.doi.org/10.2139/ssrn.626183

[10]We must not forget that the haze is not entirely manmade. Forest fires are a natural phenomenon.

only liable if the fire was manmade? Further, should the landowner be liable, or should the land user (owners often lease out their land) be liable?

Finally (for this chapter at least), there is also the issue of accuracy of data. The cost-sharing rule requires robust estimation of haze damages that countries can agree on. Since higher estimations mean a greater share of the abatement costs, countries have incentives to argue for lower estimated damages. This could compromise the accuracy of the data.

There are other issues that prevent the direct application of cooperative game theory to the SEA transboundary haze pollution problem. However, the point of this chapter is not to provide a laundry list of limitations. Instead, the point that is being made is a broader one — the limitations of a game theoretic solution extend beyond economics and a comprehensive solution must therefore also extend beyond economic analysis.

The light through the haze

An economist must be an optimist in the study of the dismal science to be of any use to the world. In the spirit of optimistic toiling in the field of the dismal science, this last section attempts to show how some of the limitations of the previous section may be overcome.

Free-riding is an economic phenomenon and there are economic solutions to it. Economists have been working on increasingly complex rules of distribution to account for the free-riding incentive. The Almost Ideal Sharing Scheme (AISS) developed by Eyckmans and Finus in 2004[11] is a case in point. Again, we will not go into the mathematics of it. Suffice to say, the AISS presents a possibility of creating a stable cooperation despite the incentive to free-ride.

Fairness might seem like a tricky issue at first glance. However, what we do know is that while people are concerned about fairness, what is perceived as fair is malleable. Paying the polluter would not seem as unfair if the polluter is seen to be putting in effort or incurring costs to

---

[11] Eyckmans, J. and Finus, M. (2004). An Almost Ideal Sharing Scheme for coalition games with externalities. CLIMNEG Working Paper 62, K. U. Leuven, Centrum voor Economische Studiën: Leuven.

curb the haze. Behavioral studies have shown that people are prepared to pay more for a good/service if they know that the cost of the good/service had increased.[12] Similarly, people will likely be willing to pay the polluter if the costs of haze abatement are made known. Additionally, framing effects could be put into practice. 'Aiding the haze prevention efforts' does not inspire the same sense of unfairness as 'paying the polluter'.

Feasibility is only an issue if one considers the current Indonesian government's resources and manpower as the constraints on how much can be directed to haze abatement. However, the required resources can and should be included in the cost of abatement to be shared by ASEAN. In that light, it should eventually be feasible to monitor and enforce whatever regulations eventually take place. The process of accounting for costs must be a transparent one, of course. While that may prove to be tricky (e.g., accounting for the cost of hiring a ranger whose job includes both monitoring the forests for fires (which clearly is a part of haze abatement) and tagging endangered species (which clearly is not) will require his wages to be split into a haze abatement component and a non-haze abatement component), it is not impossible. However, the topic of cost accounting is not the focus of this chapter.

Finally, there are established and accepted means of estimating damages in the economics literature, particularly from valuation techniques in the field of cost-benefit analysis. Environmental economist Quah once estimated that the cost of 1997 haze to Singapore was 163 million US dollars.[13] Similar studies could be done for the other ASEAN countries.

Conclusion

In conclusion, this author is hopeful that applications from game theory can eventually translate into a workable solution for the SEA transboundary pollution. If more complex issues like climate change, which involves

---

[12]Kahneman, D., Knetsch, J. L. and Thaler, R. (1986). Fairness and the assumptions of economics, *Journal of Business*, 59 (4): 285–300.

[13]Quah, E. and Tan, T. T. (2015). When the haze doesn't go away. *The Straits Times*. Singapore Press Holdings.

even more parties, can be tackled, there is little reason to believe that the SEA transboundary haze problem cannot be. The most concrete and immediate steps that should be taken are to conduct studies to obtain robust estimates of both the damages that the transboundary haze causes to individual countries, and the costs of haze abatement. Good data form the basis without which little else can proceed and there is already much to be done here.

20

Singapore's New Transboundary Haze Pollution Act: Can It Really Work to Prevent Smoke Pollution from Indonesia?

Alan Khee-Jin Tan

Professor, Faculty of Law, National University of Singapore
alantan@nus.edu.sg

Overview

On 25 September 2014, Singapore's new Transboundary Haze Pollution Act ('the Act') came into effect.[1] The Act radically creates an extra-territorial liability regime for entities such as plantation companies engaging in setting fires abroad that cause transboundary smoke or 'haze' pollution felt in Singapore. Indeed, the Act is the first of its kind in the world to prescribe extra-territorial liability for transboundary smoke pollution arising from fires deliberately set in another country.

The Act's enactment can be traced to the serious haze pollution origi-nating from Indonesia that hit Singapore in June 2013. For several days in that month, haze pollution in Singapore resulting from the massive forest and land fires in Indonesia reached an unprecedented high.[2] Prior

---

[1] Transboundary Haze Pollution Act (hereinafter 'the Act'), Statutes of the Republic of Singapore, Singapore Statutes Online, http://statutes.agc.gov.sg/aol/download/0/0/pdf/binary File/pdfFile.pdf?CompId:e2031db7-7071-4016-9060-80de762953ef accessed 1 June 2016.

[2] The air pollution measurement index used in Singapore (the PSI — Pollutant Standards Index) hit a historical high of 401 on 21 June 2013, well over the 'hazardous' level of 300

to 2013, the most serious haze pollution had occurred in 1997, the year Indonesia was struck by the Asian financial crisis and the ensuing political upheavals that ousted President Suharto. I had argued then for Indonesian state responsibility for failing to control the fires.[3] Since that time, the fires and transboundary haze have continued to be serious regular occurrences that affect principally Indonesia's three neighbours: Malaysia, Singapore and Brunei. The most recent serious haze pollution was in September 2015, when air pollution levels rose into the unhealthy range for several days.

Satellite images of burnings in Indonesia reveal that the fires occur principally within large oil palm and *Acacia* plantation concessions (*Acacia* is used for the pulp and paper industry).[4] Indeed, over half of the satellite 'hotspots' during the 2013 fires were recorded within oil palm and *Acacia* concessions.[5] In Indonesia, fire has been used for generations to clear land for agriculture, both by smallholders and large companies. Yet, the plantation interests have consistently denied their role in the fires, placing the blame instead on small-scale farmers and local communities living near or within their concessions.[6] On its part, the Indonesian government promised tough action against the perpetrators, but as in previous years, failed to tackle the problem effectively.

---

and far exceeding the worst level of 226 recorded in 1997, Shukman, D. (2013, June 21). Singapore haze hits record high from Indonesia fires. *BBC News*. Retrieved http://www.bbc.com/news/world-asia-22998592, accessed 1 June 2016. In fact, the PSI breached the 300 level on four successive days in June 2013, causing widespread public alarm. In September 2015, the PSI reached a year-high of 341.

[3] Tan A. K-J. (1999). Forest fires of Indonesia: State responsibility and international liability, *International and Comparative Law Quarterly*, 48: 826.

[4] The World Resources Institute (WRI) estimates that over half of fire alerts in Riau Province (one of the worst-affected areas) occurring between 12 to 23 June 2013 were within concession areas, Neo, C. C. (2013, June 28). Haze finger-pointing: Time for companies to show and tell. *TODAY*. Retrieved http://www.todayonline.com/commentary/haze-finger-pointing-time-companies-show-and-tell, accessed 1 June 2016.

[5] Gaveau, D. (2014). New maps reveal more complex picture of Sumatran fires, Centre for International Forestry Research. Retrieved http://blog.cifor.org/23479/new-maps-reveal-more-complex-picture-of-sumatran-fires#.VeRzfkP2NyR, accessed 1 June 2016.

[6] Neo, note 4.

In the aftermath of the June 2013 fires, during a hastily-convened emergency meeting of regional environment ministers, Singapore openly raised the prospect of Indonesia sharing its plantation concession maps that would disclose the geographical coordinates of offending planta-tions.[7] The aim was to identify precisely the locations of the fires and to link them to the companies that control the land. When placed alongside satellite 'hotspot' images, it was thought that the maps could provide the necessary proof for Indonesia to take enforcement action against the plan-tations as well as to anticipate fires during future burning seasons.

In essence, Singapore was responding to Indonesian officials' fre-quent allegations (no doubt to deflect criticisms of inaction on their own part) that many of the plantations were foreign-owned or -controlled, with several being based in Singapore and Malaysia or even listed on these countries' stock exchanges.[8] Indeed, at least seven companies with exten-sive oil palm and pulp and paper operations in Indonesia have a presence in Singapore.[9] The Singapore government thus argued that it needed the maps to verify who the concessionaires were so that it could pursue action against any offending Singapore-linked entity. Such action was not straightforward as it entailed adopting extra-territorial legislation to target offenders for their conduct abroad. This was something Singapore sig-naled it was prepared to consider.

The Indonesian government, however, has refused to share the relevant map information. Indeed, its Environment Minister even raised legal obsta-cles and insisted that the country's freedom of information laws prohibited the public disclosure of such information.[10] Ironically, this stance was shared by Malaysia, whose Environment Minister claimed that land mat-ters fell within the authority of individual states in Malaysia, as opposed to

---

[7]The meeting was the 15th Meeting of the Sub-Regional Ministerial Steering Committee (MSC) on Transboundary Haze Pollution held in Kuala Lumpur in July 2013.

[8]Bland, B. (2013, June 23). Indonesian fires highlight weak governance and corruption. *Financial Times*. Retrieved http://www.ft.com/cms/s/0/a6d8c050-dbf5-11e2-a861-00144 feab7de.html, accessed 1 June 2016.

[9]These include Wilmar, Asia Pulp and Paper, Golden Agri-Resources and APRIL.

[10]Hussain Z. (2013, July 19). Jakarta's information law forbids sharing of maps. *Straits Times*. Retrieved http://news.asiaone.com/News/Haze/Story/A1Story20130719-438487. html, accessed 1 June 2016.

the federal government.[11] Due to this refusal, the three countries could only agree to share maps in a limited manner between governments, and on a "case-by-case basis."[12]

To date, no maps appear to have been shared. In fact, Indonesia has reaffirmed its refusal to provide the maps.[13] The likely reason is that the Indonesian government itself may not possess accurate maps and information on the plantations. This explains why it is currently pursuing a nation-wide 'One Map Initiative' that aims to chart a single all-encompassing map to rationalize information on forest licenses, agricultural concessions and other land use claims.[14] If successful, this overdue national map project can be expected to resolve overlapping land use claims and the resulting uncertainty over who bears responsibility for setting fires. For now, the problem is the project's slow implementation; estimates are that it may take another three years to complete.[15] Until then, it is unlikely that Indonesia will be willing to share any of its current maps.

The Transboundary Haze Pollution Act and its implications

In Singapore, public frustration with the recurring transboundary haze, particularly the severe episode of June 2013, led the government to consider extra-territorial legislation to impose criminal and civil liability on agri-business companies involved in using fires outside Singapore, whether these be Singapore-linked or otherwise. The resort to domestic law is also driven by the realization that action against Indonesia before

---

[11] Ibid.

[12] Ibid.

[13] Soeriaatmadja, W. (2015, July 29). Ministers agree to share hot spot info. *Straits Times*. Retrieved http://www.straitstimes.com/asia/se-asia/ministers-agree-to-share-hot-spot-info, accessed 31 August 2015. The three countries agreed on sharing satellite 'hotspot' images, but failed to agree on the maps. Hence, even if the fires can be pinpointed, it would still be uncertain whose lands they were on.

[14] Sizer, N. (2013, July 17). Haze risk will remain high. *Jakarta Post*. Retrieved http://www.thejakartapost.com/news/2013/07/17/haze-risk-will-remain-high.html, accessed 1 June 2016.

[15] Ibid.

international tribunals is unrealistic.[16] At the same time, the domestic action is meant to rebut the Indonesian argument that the victim states fail to look to their own companies first. Thus, on 19 February 2014, a proposed Transboundary Haze Pollution Bill was announced for public consultation. By early August 2014, the Bill had been debated in Parliament and passed as an Act.

On 25 September 2014, the Transboundary Haze Pollution Act came into effect, laying out potential extra-territorial criminal and civil liability for offending companies. Under the Act, a convicted entity that engages in conduct, or engages in conduct that condones any conduct by another entity or individual which causes or contributes to any haze pollution in Singapore (or the entity that participates in the management of a second entity that owns or occupies land and engages in the relevant conduct) can face a fine not exceeding 100,000 Singapore dollars (about 80,000 US dollars) for every day or part thereof that there is haze pollution in Singapore.[17]

If the entity has failed to comply with any preventive measures notice, there can be an additional fine of up to 50,000 US dollars (40,000 US dollars) for every day or part thereof that the entity fails to comply with the notice.[18] Overall, there is a maximum aggregate fine of 2 million Singapore dollars (1.6 million US dollars). In addition, a civil liability regime is prescribed. Affected parties (such as hotels suffering booking losses) may thus bring civil suits against entities causing or contributing to haze pollution in Singapore.[19] The civil damages recoverable are theoretically unlimited and will be determined by the court based on evidence of personal injury or physical damage to property or economic loss (including loss of profits).

We next analyze aspects of the Act that may prove challenging in the context of adversarial action (whether civil or criminal) against relevant

---

[16] A state's appearance before an international court or tribunal cannot be compelled. In other words, it would require Indonesia's consent, something that is unrealistic to expect, Tan, note 3.

[17] Transboundary Haze Pollution Act, ss 5(1), 5(2), 5(3) and 5(4).

[18] Transboundary Haze Pollution Act, ss 5(2)(b) and 5(4)(b).

[19] Transboundary Haze Pollution Act, s 6.

offenders before the Singapore courts. Arising from the September 2015 haze pollution, and with a view to possible prosecution, the Singapore authorities have issued orders to six Indonesian companies to provide information on their suspected burning activities.[20] These companies include PT Rimba Hutani Mas, PT Sebangun Bumi Andalas Wood Industries, PT Bumi Sriwijaya Sentosa, PT Wachyuni Mandira and PT Bumi Mekar Hijau. Several of these companies — including PT Bumi Mekar Hijau — are subsidiaries of Asia Pulp and Paper (APP), a Singapore-based pulp and paper company that is linked to the Sinar Mas conglomerate in Indonesia. More information on the ongoing investigations is provided at the end of this chapter.

Proof and legal presumptions

The Act's centerpiece is a regime of multiple legal presumptions. First, it provides that if there are maps which show that any land is owned or occupied by a company, it shall be presumed that that company owns or occupies that land.[21] The maps can be procured from a variety of sources — the Act contemplates these to include any foreign government, any department or instrumentality of the government of a foreign state (presumably, this includes the provincial, regency or village authorities) or any person who can be legally compelled to furnish any maps.[22] This would mean that a company or entity operating a concession can be compelled to furnish its own maps.[23]

Second, if there is serious haze pollution in Singapore and satellite and other meteorological evidence show that at or about that time, there is a land or forest fire on any land causing smoke that is moving in the

---

[20] Straits Times. (2015, September 30). NEA asks another Indonesian firm to take measures to mitigate haze conditions. *Straits Times*. Retrieved http://www.straitstimes.com/singapore/environment/nea-asks-another-indonesian-firm-to-take-measures-to-mitigate-haze-conditions, accessed 1 June 2016.

[21] Transboundary Haze Pollution Act, s 8(4).

[22] Ibid.

[23] It will be difficult to compel a foreign entity that has no presence at all in Singapore. This is why the Act, despite its extra-territorial ambition, is likely to be effective only against entities that have a presence in Singapore.

direction of Singapore, it shall be presumed that there is haze pollution in Singapore involving smoke resulting from *that* land or forest fire.[24] Crucially, this is so even if there may be fires in other or adjacent areas at or about the same time.[25]

Third, it shall be presumed that the company that owns or occupies the land in question has engaged in conduct, or engaged in conduct that condones any conduct by another, which caused or contributed to that haze pollution in Singapore.[26]

The entity or company concerned can deny each of these presumptions but will bear the heavy burden of proving the contrary. Further, the Act extends liability to any entity that participates in the management or operational affairs of another (second) entity, exercises decision making control over the latter's business decision pertaining to land that it (the second entity) owns or occupies outside Singapore, or exercises control over the second entity at a level comparable to that exercised by a manager of that entity.[27] This is designed to target parent or holding companies that have subsidiaries or related entities that are the owners or occupiers of land and that engage in offending conduct on the ground.

As regards the series of presumptions, what is likely to be challenging is the very first presumption. Even if maps can be obtained from any of the relevant sources, the assumption here would be that these maps are authoritative, updated and accurate. However, different 'official' maps or versions typically exist in Indonesia, depending on their source. This is because the different agencies frequently issue overlapping land use rights and their respective maps are often contradictory in identifying the boundaries and the respective owners or occupiers.[28] In short, there would be uncertainty if the prosecution relies on maps that have discrepancies and are contested.

Of course, the prosecution in Singapore could avoid these problems by compelling the companies to furnish their own maps. In this manner,

---

[24] Transboundary Haze Pollution Act, s 8(1).
[25] Ibid.
[26] Transboundary Haze Pollution Act, s 8(2).
[27] Transboundary Haze Pollution Act, ss 3 and 8(3).
[28] Sizer, note 14.

any area stated as within a company's map boundaries must be presumed to be owned or occupied by it. Requirements to lodge or provide maps may also be imposed as part of stock exchange disclosures or company registration requirements in Singapore.

The effect of all this is to put the company in the invidious position of either acknowledging or denying the accuracy of its own maps. As a result, this could encourage the companies to obfuscate or qualify their own maps further by marking vast areas as being 'contested' or indeterminate. This situation could well arise if the companies are operating alongside local communities that dispute the boundaries and thus encroach into the companies' lands.[29]

One typical scenario is where the company operates on land that is actually smaller in area than the official concession boundaries granted by the relevant authorities.[30] Hence, there would be significant discrepancies between the official maps and those being used by the companies themselves. There will thus be confusion over fires on land thought to be within the limits of the concessions but that are actually outside areas that the companies fully control and are directly developing.[31] The implication here is that the fires would have been started by third parties.

In such situations, the Act would still treat these companies as the rightful owners or occupants, given the existence of official maps that attest to this fact. The companies would then have to raise the defense that the offending conduct was by another person acting without their knowledge or consent, or contrary to their wishes or instructions.[32] The Act stipulates that such a defense is unavailable if the other person is an employee or agent of the companies, or is engaged, directly or indirectly, by the companies to carry out any work on land that they own or occupy, or is any person who has a customary right as regards the land and who has an agreement or arrangement with the companies relating to farming or forestry operations in respect of that land.[33] Hence, the burden of

---

[29] Encroachment by other land users who start fires is common, Gaveau, note 5.

[30] Sizer, note 14.

[31] Sizer, note 14.

[32] Transboundary Haze Pollution Act, s 7(2).

[33] Transboundary Haze Pollution Act, s 7(2)(a)-(c).

disassociating itself from the third party's offending conduct is a heavy one for the company.

As stated earlier, the Act provides that maps can be furnished by or obtained from various sources, including any person specified in the Act's Schedule through any means specified in that Schedule. Here, the Act contemplates that "specified persons and means" can be very wide in scope, including various individuals and websites.[34] This contemplates non-governmental sources such as environmental groups and foreign entities. In this regard, there are concerns over the accuracy of materials obtained from such a possibly wide variety of persons and means. It appears that the only way for a company to guard against potentially hostile or unreliable sources is for it to maintain and update its own maps proactively *and* to furnish these voluntarily to the authorities in Singapore.

Overall, the first presumption in the Act is powerful but not necessarily watertight. Prosecutors are likely to proceed cautiously, particularly with the first test-case(s) as there will be enormous public expectation in Singapore for a successful conviction. On its part, the adjudicating court can be expected to be cautious as regards maps that are contested, outdated or unreliable. As such, the prosecution is likely to charge a company only if there is accompanying independent evidence to support the first presumption. These other evidence would include alternative maps (that the courts may rule to be more reliable), documentary proof of the company's prior dealings with the land in question, or direct employee or local resident testimonies of the company's offending activities on that land.

Rebutting the presumptions

While not impossible, rebutting the presumptions will be a difficult task for the company. For the first presumption, the company will have to prove that contrary to the prosecution's maps, it does not own or occupy the land in question. If the maps have been obtained from independent sources, the company may conceivably reveal its own maps and argue that these are more authoritative and accurate. If, however, the prosecution has

---

[34] Transboundary Haze Pollution Act, s 8(4).

obtained the maps from the company itself, the presumption is unlikely to be displaced.

It is conceivable, however, that the maps (even the ones the company itself relies on) can subsequently be claimed to be inaccurate. Hence, a particular tract of land with alleged fires on it may actually turn out not to be owned or occupied by the company.[35] Here, it is possible that the company may have been mistaken all along about its ownership or occupation of the land (this also raises the general defense of mistake of fact, see below).

A more likely scenario would be where the company never regarded a tract of land to be under its ownership or occupation (despite government maps indicating so). As stated earlier, a company's actual operations may be smaller than the theoretical concession limits granted by the government.[36] In this situation, the company will have to question the accuracy of governmental maps and to argue that it always knew or assumed that those maps did not accord with realities on the ground. In other words, it never occupied or had control of that land and never behaved to that effect. To bolster its case, the company would have to find maps or evidence from other sources that attest to its non-ownership or non-occupation, or actual ownership or occupation by third parties.

The company may also have marked out certain areas within its concession as being disputed or contested. As stated above, it is unclear if this will be sufficient to rebut the first presumption. In other words, the courts may still find the first presumption to be satisfied and to hold that the company does own or occupy the land. The critical issue here would shift to the company's defense (see below) that the fires had been started by third parties on disputed lands over which it had no actual control, even if government maps did show that the areas came within its concession.

If the company fails to rebut the first presumption, the matter moves on to the second presumption. This will be extremely difficult to rebut, unless the company is able to show that the satellite and meteorological evidence tendered by the prosecution is unreliable with regard to either or both of the following respects: (i) that a fire is indeed occurring on land

---

[35] Sizer, note 14.
[36] Sizer, note 14.

owned or occupied by it; or (ii) that any smoke resulting from that fire is moving in the direction of Singapore.

On (i), it will have to tender evidence to show that there is *actually* no fire occurring on land owned or occupied by it (or, by operation of the first presumption, presumed to be owned or occupied by it). In other words, it will have to produce evidence such as aerial photography or eyewitness testimonies to disprove the satellite hotspots and to show, for instance, that there are 'false-positive' problems with the hotspot images.[37]

On (ii), the company will bear the burden of showing that any smoke resulting from the fire on land owned or occupied by it is not moving in the direction of Singapore. At the same time, it may seek to prove that the smoke pollution found occurring in Singapore does not arise from the fires on land that it owns or occupies. This will be difficult given that the Act has rendered it irrelevant whether there may be, at or about the same time, any land or forest or other fire on any other land situated outside Singapore, whether or not adjacent to the company's land. Also, as is obvious, it is not as if the haze pollution found in Singapore can be chemically analyzed and traced to a particular source (the way industrial pollutants possibly can).

To rebut the third presumption, the company will have to prove that it has not engaged in conduct, or engaged in conduct that condones any conduct by another entity, which caused or contributed to haze pollution. Here, the conduct covers both acts and omissions (e.g., a failure to control or stop). Under the Act, there is both criminal and civil liability for an entity that participates in the management or operational affairs of another (second) entity, exercises decision making control over the latter's business decision pertaining to land that it (the second entity) owns or occupies outside Singapore, or exercises control over the second entity at a level comparable to that exercised by a manager of that entity.[38]

Again, it would be difficult for the first company to rebut this presumption. Presumably, it can show that neither it nor its employees have

---

[37] 'False-positives' refers to errors that cause hot-spots to show up even where there is no fire. The reason could be fires or heated ground in nearby areas or cloud cover that affects the accuracy of monitoring.

[38] Transboundary Haze Pollution Act, ss 5(3) and 6(2).

engaged in such conduct. In addition, it would have to show that it has not condoned such conduct. Here, one must take into account the realities of plantation operations in Indonesia. In some places, it is common (and local law may so require) that companies must sub-contract certain tasks to local villagers or co-operatives.[39] In such situations, the company would have to show that it explicitly issued instructions to its sub-contractors *not* to use fire. If it has done so, it may well succeed in showing that it has not condoned the use of fires.

However, the Singapore courts may construe the word 'condone' to mean that the company knew or should have known that the sub-contractors would or might use fires to clear land, since this is a commonly practiced method. To disprove the concept of 'condone' here, the company would likely have to tender evidence that it had done one or more of the following: (i) explicitly warned the sub-contractors that if they used fire, they would be breaching their agreement and that their services would be terminated; (ii) practiced a consistent policy of terminating the services of sub-contractors who used fires; (iii) provided and trained the sub-contractors to use alternative means such as mechanical devices (bulldozers) to clear land; (iv) warned them that they would be breaching Indonesia's laws and that they would be reported to the police or relevant authorities; and (v) reported such transgressors to the authorities.

In other words, the company would have to show that it had taken proactive steps to disassociate itself from the sub-contractors' use of fire, and that these parties had gone ahead to ignore its explicit instructions not to use fire. Even if all these could be proven, the courts could still take the position that the company must assume responsibility for the conduct of its sub-contractors, either because these are to be treated as agents or the equivalent of employees.[40] At the same time, the courts are unlikely to entertain any claim that the areas in question are too large for mechanical bulldozers, or that using fire is the only realistic method for clearing land.

---

[39] Chua, G. and Nazeer, Z. (2014, February 23). 'Yes' to law on transboundary haze. *Straits Times*. Retrieved http://news.asiaone.com/news/singapore/yes-law-transboundary-haze, accessed 1 June 2016.

[40] Transboundary Haze Pollution Act, s 7(2)(a).

The applicable defenses

Even where all the presumptions are satisfied, the company may still plead certain defenses. Strict liability offences such as those prescribed under the Act may attract two types of defenses — those provided explicitly under the Act itself, and general defenses applicable to all criminal prosecutions under Singapore law.

Defenses under the act

The explicit defenses under the Act are as follows (these apply to both the criminal and civil actions)[41]:

7.— (1) It shall be a defence to a prosecution for an offence under section 5(1) or (3), and to a civil claim for a breach of duty under section 6(1) or (2), if the accused or defendant (as the case may be) proves, on a balance of probabilities, that the haze pollution in Singapore was caused solely by —

(a) a grave natural disaster or phenomenon; or
(b) an act of war.

(2) It shall also be a defence to a prosecution for an offence under section 5(1) for engaging in conduct which causes or contributes to any haze pollution in Singapore, and to a civil claim for a breach of duty under section 6(1) not to engage in conduct which causes or contributes to any haze pollution in Singapore, if the accused or defendant (as the case may be) proves, on a balance of probabilities, that the conduct which caused or contributed to the haze pollution in Singapore was by another person acting without the accused's or defendant's knowledge or consent, or contrary to the accused's or defendant's wishes or instructions; but that other person cannot be —

(a) any employee or agent of the accused or defendant (as the case may be);

---

[41] Transboundary Haze Pollution Act, s 7.

(b) any person engaged, directly or indirectly, by the accused or defendant (as the case may be) to carry out any work on the land owned or occupied by the accused or defendant, and any of that person's employees; or

(c) any person who has a customary right under the law of a foreign State or territory outside Singapore as regards the land in that foreign State or territory and with whom the accused or defendant (as the case may be) has an agreement or arrangement, which agreement or arrangement relates to any farming operations or forestry operations to be carried out by any person in respect of that land.

(3) It shall also be a defence to a prosecution for an offence under section 5(1) for engaging in conduct which condones any conduct by another entity or individual which causes or contributes to any haze pollution in Singapore, and to a civil claim for a breach of duty under section 6(1) not to engage in conduct condoning any conduct by another entity or individual which causes or contributes to any haze pollution in Singapore, if the accused or defendant (as the case may be) proves, on a balance of probabilities, that —

(a) the accused or defendant took all such measures as is (or was at the material time) reasonable to prevent such conduct by the other entity or individual; and

(b) if the conduct by the other entity or individual already occurred, the accused or defendant took all such measures as is (or was at the material time) reasonable to stop that conduct from continuing or to substantially reduce the detriment or potential detriment to the environment in Singapore or its use or other environmental value, or the degradation or potential degradation to the environment in Singapore, due to the other entity's or individual's conduct.

(4) It shall also be a defence to a prosecution for an offence under section 5(3), or a civil claim for a breach of duty under section 6(2), if the accused or defendant (as the case may be) proves, on a balance of probabilities, that the conduct which caused or contributed to the haze pollution in Singapore was by another person acting without the knowledge or consent of the accused or defendant and the second entity referred to in section

5(3) or 6(2), or contrary to the wishes or instructions of the accused or defendant and that second entity; but that other person cannot be —

(a) any employee or agent of the accused or defendant (as the case may be) or of the second entity referred to in section 5(3) or 6(2);
(b) any person engaged, directly or indirectly, by the accused or defendant (as the case may be) or by the second entity referred to in section 5(3) or 6(2), to carry out any work on the land owned or occupied by the second entity, and any of that person's employees; or
(c) any person who has a customary right under the law of a foreign State or territory outside Singapore as regards the land in that foreign State or territory, and with whom the accused or defendant (as the case may be) or the second entity has an agreement or arrangement, which agreement or arrangement relates to any farming operations or forestry operations to be carried out by any person in respect of that land.

The 'grave natural disaster or phenomenon' and 'act of war' defenses are straightforward. As with all defenses, the company has the burden to prove them on a balance of probabilities. However, the challenge here is to show that the haze pollution was caused *solely* by a grave natural disaster or phenomenon or an act of war. In other words, the company has the burden to show that it had no contributory role whatsoever. For instance, in the case of a lightning strike (a natural phenomenon) on peat-rich land owned or occupied by the company, it must show that the lightning was the sole cause, and that no other human factors attributable to it were present.

Indeed, the fires and haze pollution could result from a combination of natural and third party man-made causes. If there is evidence of arson or fires being set by local communities encroaching into the company's land, it appears that the 'natural disaster or phenomenon' defense will be disqualified (due to the word "solely"). The company must then move on to establish the 'third party act' defense independently. It must show that the conduct which caused or contributed to the haze pollution in Singapore was by another person acting without its knowledge or consent, or contrary to its wishes or instructions.

Here, issues may arise with sub-contractors and third party encroachers.[42] For sub-contractors, the courts are likely to hold these to be employees or agents, or persons "engaged directly or indirectly" by the company.[43] The major difficulty arises with third party encroachers, e.g., local communities who either encroach into the company's land to set off fires or whose use of fires outside the land spreads into the company's areas.[44] If the company can prove that the fires have been set off by these parties without its knowledge or consent, or contrary to its wishes or instructions, the defense is likely to be made out. However, the defense is not available if the encroachers are local communities with customary rights over the land and with whom the company had an agreement or arrangement relating to farming or forestry operations.[45]

As is typical in Indonesia, the plantation companies may have entered into agreements to set aside adjacent tracts of land for local communities that they had displaced. These communities would then cultivate crops, some of which may even be processed by the companies. Such arrangements are common either because the companies do this out of goodwill or because they form part of the terms of obtaining the concessions. In either case, the companies are expected to accommodate existing communities living in or near the concession areas. The Act thus presumes that the companies must take responsibility for the actions of these communities.

The more responsible companies typically attempt to support the local communities' livelihood and to educate them on the perils of using fires. If, however, these communities persist in using fires, this might suggest that the companies had knowledge of their conduct, or that an agreement or arrangement to use that land in the relevant manner had been entered into. If so, it is unlikely that the Section 7(2) defense will succeed.

There could also be situations where a company may simply have tolerated the presence of local communities within or adjacent to their concession areas, without explicitly agreeing to host them. In an ideal

---

[42] Chua and Nazeer, note 39.
[43] Transboundary Haze Pollution Act, s 7(2)(a)-(b).
[44] Gaveau, note 5 and Sizer, note 14.
[45] Transboundary Haze Pollution Act, ss 7(2)(c) and 7(4)(c).

situation, the company should have evicted such third parties altogether. In Indonesia, however, such eviction is unrealistic, given the entrenched uncertainties in land use tenure and the sensitivity of evicting local communities forcibly. In such circumstances, it would be a harsh court that holds the company to have had an agreement or arrangement with these encroachers.

Another relevant defense arises under Section 7(3), where the relevant charge against the company would relate to engaging in conduct that condones the act of a third party that causes or contributes to haze pollution. The company's defense here is to show that it took all such measures as was reasonable to prevent such conduct by the third party. If the conduct had already occurred, the company must show that it took all reasonable measures to stop that conduct from continuing or to substantially reduce the detriment or potential detriment to Singapore. Here, the company would probably stand a good chance of arguing that it has not condoned third party conduct if it can show that it had systematically educated the local communities on sustainable farming practices, but that they had persisted in using fires despite the company's best efforts to prevent this.

On the question of local communities, complications may also arise from certain provisions of Indonesian law that clearly condone such communities' use of fire to clear land. For instance, the 2009 Law for Protection and Management of the Environment (Law No. 32 of 2009) provides in its Article 69, paragraph 1(h) that:

> Every person is prohibited from ... (h) ... conducting the clearing of land using burning.

Paragraph 2 of the same article clarifies that paragraph 1(h) accords the highest attention to "local wisdom" in the individual regions. In turn, Law No. 32 of 2009's Explanation section (a common interpretation tool in Indonesian legislation) provides that:

> The local wisdom meant in this provision is the conducting of land burning with a maximum land area of 2 hectares per head of family for the cultivation of local varieties of plants and that is ringed by fire breaks to prevent the spread of fire to surrounding areas.

As such, Law No. 32 of 2009 actually provides explicitly for villagers to conduct land clearing by fire for the purpose of agricultural livelihood. The provision for "2 hectares per head of family" refers to the area that land settlers (typically, transmigrants that are moved from other parts of the Indonesian archipelago) are given to support their livelihood.[46] This clearly points to burning practices being a reality in local life.

Consequently, prosecuting a company under the Act could be complicated by the presence of local communities using fires legitimately to clear land and the likelihood that such fires could spread into neighboring plantations. The fact that Indonesian law actually provides an allowance for local communities to use fires could be used by the plantation companies as a defense, provided they show that they did not condone the use of such methods. The companies may even seek to show that as part of their 'due diligence' conduct, they had actively provided fire or fuel breaks like trenches to prevent the spread of fires. Alternatively, they could seek to show that the villagers themselves did not construct the firebreaks — it is unclear from Law No. 32 of 2009 whose responsibility it is to provide those firebreaks; presumably, the responsibility lies with the villagers themselves.

Interestingly, there is nothing in the Act to suggest that the villagers or local communities themselves cannot be liable for prosecution or civil action in Singapore. Of course, it would be unrealistic and impractical to pursue action against individual villagers or farmers. However, to the extent that these parties commonly belong to village or sub-district co-operatives in Indonesia, it is not inconceivable that these entities can be held liable under the Act. After all, the Act defines an "entity" to mean "sole proprietorship, partnership, corporation or other body of persons, whether corporate or unincorporate".[47] A village co-operative would, at the minimum, be an unincorporated body of persons. However, it does mean that any prosecution under the Act will encounter the problem that the local communities' use of fires to clear land is actually contemplated and allowed under Indonesian law!

---

[46] Hardjono, J. M. (1977). *Transmigration in Indonesia*, Oxford University Press, Kuala Lumpur, 40.
[47] Transboundary Haze Pollution Act, s 2.

The other party that is outside the reach of the Act appears to be governments which would enjoy sovereign immunity in any event. That said, sovereign immunity is not absolute, as international law principles provide for a 'commercial activity' exception that allows governments to be subject to a foreign court's jurisdiction for acts that are private or commercial (*jure gestionis*) as opposed to governmental or sovereign in nature (jure imperii).[48] In any event, local governments at the municipal or city levels are not generally entitled to sovereign immunity.[49] Even so, it is highly unlikely that the Act will be used by Singapore against any Indonesian government unit, be it central or local, as this would be politically untenable. The same cannot be said for Indonesian government-linked entities that engage in commercial or business activities. These would be fully open to criminal and civil action under the Act.

General defenses under criminal law

Under Singapore criminal law, the general defenses provided under Part IV of the Penal Code shall apply to all criminal provisions in any statute. For strict liability offences such as those prescribed in the Act, the defense of mistake of fact under section 79 of the Penal Code shall generally be applicable. Section 79 reads:

> Nothing is an offence which is done by any person who is justified by law, or who by reason of a mistake of fact and not by reason of a mistake of law in good faith believes himself to be justified by law, in doing it.

The mistake must be one held in good faith. This concept is, in turn, defined by Section 52 of the Penal Code:

> Nothing is said to be done or believed in good faith which is done or believed without due care and attention.

---

[48] Yang, X. (2012). *State Immunity in International Law*, Cambridge University Press, 207.

[49] Fox, H. and Webb, P. (2013). *The Law of State Immunity*, Oxford University Press, 350.

Furthermore, section 32 states:

> In every part of this Code, except where a contrary intention appears from the context, words which refer to acts done extend also to illegal omissions.

The key in proving a defense of mistake of fact lies in showing that the accused committed the mistake despite having taken due care and attention. This defense could be relevant, for instance, if the company had good reason to believe that certain tracts of land did not come within its ownership or occupation (although in fact, governmental maps showed they *did*).[50] The company could thus show that its failure to control (i.e., an omission amounting to an act) the use of fires by third parties on that land was borne out of a mistake of fact held in good faith. This will be difficult because the company would have to show that it had taken all reasonable steps to check with the relevant authorities and that it had come to the reasonable conclusion that it had no ownership or occupation rights over that land.

The other more general defense that could apply to strict liability offences is the common law defense of 'due diligence'. Here, the company would have to show that it had taken all reasonable steps in the circumstances to prevent and control the use of fires on lands that it owns or occupies. Of course, this is closely related to the explicit statutory defenses provided under the Act discussed above. For instance, the company would have to show that it had taken all reasonable steps to provide its sub-contractors with alternative means of land clearing and to prohibit the use of fires.

On encroachers, it would have to show that it took all reasonable steps to prevent encroachment (this is difficult if the boundaries are disputed) and to prohibit the use of fires. Other 'due diligence' steps the company could demonstrate include: adopting and practicing sustainable forestry and planting practices, educating employees, sub-contractors and local communities on the danger of using fires, reporting transgressors to public authorities, and enforcing a 'zero burning' policy consistently throughout its concessions.

---

[50] See above, text to note 30.

In general, it would be difficult to succeed on this 'due diligence' defense, even if the courts gave the company a wide measure of latitude for the size of its concessions. It boils down to whether the courts are prepared to accept that the company has done all that it could reasonably have done in the circumstances to prevent the use of fires, including in attempting to control the conduct of its employees and sub-contractors, and more problematically, that of local communities living within or adjacent to areas under its ownership or occupation.

Additional evidence needed beyond maps, satellite images and legal presumptions

Even with the aid of the legal presumptions discussed above, any potential prosecution before the courts must grapple with critical questions of evidence and burden of proof. As discussed above, it is conceivable that a 'strict liability' regime premised on a series of legal presumptions that throw the burden of disproving certain facts onto the companies themselves can be employed under the victim states' own laws. However, the robustness and fairness of such a system must depend on the accuracy of the maps and the availability of supporting evidence. The maps' accuracy, as it is now clear, cannot be assumed.

It would seem that even the much-vaunted satellite identification technology is not without its issues. While generally regarded as reliable, the accuracy of satellite tracking can be compromised by thick cloud cover and other weather-related factors. 'Hotspots' showing up on satellite images may not necessarily correlate to actual fires on the ground. In fact, a fire can be represented by a number of adjacent hotspots, and it is not necessarily the case that one fire corresponds to one hotspot.

In addition, 'false positives' regularly show up on satellite images — extremely hot soil surface, particularly in peat-rich areas, or even the sun's glint, can show up as hotspots even where there is no actual fire.[51] The use of hotspots as prosecution evidence is also problematic because it is

---

[51] Johnston, L. *et al.* (2015). Indonesia's forest fires reignite, threatening protected areas and peatlands, World Resources Institute. Retrieved http://www.wri.org/blog/2015/07/indonesia%E2%80%99s-forest-fires-reignite-threatening-protected-areas-and-peatlands, accessed 1 June 2016.

highly unlikely that the accused or defendant company can produce aerial photographs or call witnesses, *after the fact*, to prove that the hotspots were not fires after all.

Due to these problems associated with maps and satellite evidence, it is likely that the prosecution will have to seek additional, independent evidence to bolster its case. This is particularly so if the relevant maps are outdated or unclear, or if the fires appear to occur in areas marked out as disputed. To begin with, the prosecution is likely to charge for fires that occur within areas indisputably owned or occupied by a company such as in the heart of the concession, as opposed to areas that are at the periphery or near contested boundaries. Even then, the prosecution would be well advised to find other evidence in the form of aerial photography that shows the sources of fires and smoke more accurately than satellite imagery can.

At the same time, on-the-ground witness testimonies would be critical in identifying where exactly the fires have been deliberately ignited (and possibly by whom). Collecting such evidence — both witness testimonies and aerial photography using low-flying reconnaissance planes — requires the explicit consent of Indonesia to enter or overfly its territory. If Indonesia's consent can be obtained through diplomatic channels, securing such evidence would bolster the prosecution case in Singapore immensely. Finally, the prosecution in Singapore may conceivably tender evidence brought before the Indonesian courts themselves if prosecution against the same companies had occurred in Indonesia. On this point, it is unclear under the Act if the principle against 'double jeopardy' applies to prevent prosecution against a company that has already been convicted in Indonesia for the same acts.

The evidence collected by independent observers in Indonesia suggests that deliberate land clearing by fire occurs predominantly in areas where land is cleared for oil palm cultivation.[52] The soaring price of palm

---

[52] Velde, B. V. (2014). Stakeholders in Jakarta map out long-term solutions to haze crisis, Centre for International Forestry Research. Retrieved http://blog.cifor.org/21463/stakeholders-in-jakarta-map-out-long-term-solutions-to-haze-crisis#.VePEPEP2NLM, accessed 1 June 2016.

oil in recent years, coupled with the generous return on investments (oil palm trees continue to be productive for some two decades) has resulted in massive clearance of land by fire for oil palm cultivation. In addition, when mature palm trees need to be replaced, clearance by fire remains the cheapest and most convenient method. For these reasons, the Singapore prosecution is likely to build, as its first test-case, a charge against an oil palm operation.

In this regard, it is important to appreciate the identities of the players on the ground. In Riau Province alone, an estimated 50 percent of total cultivated oil palm land is run by the big companies.[53] Forty percent of cultivated land is operated by small operations such as village or local communities, with the remaining 10 percent being accounted for by mid-sized companies. Forest research organizations such as the Centre for International Forestry Research (CIFOR) agree that both large-scale and small-holder operators have been alleged to use fires.[54] The scale of the fires caused by each respective camp, however, remains uncertain.

For this reason, it is important to distinguish between operators who are more responsible (or who claim to be so) and those who flagrantly use fires to clear land. Even among Singapore-based or -linked operations, there are those with huge concessions and others with smaller, mid-sized operations. Anecdotal evidence collected by the present author suggests that the transgressors are more likely to be mid-sized players with less of a reputation to protect.[55] Even assuming that the use of fires is equal among the different players, prosecuting a large-scale operator alone leaves at least half of the fires and haze problem unresolved. Consequently, obtaining the requisite intelligence and evidence — primarily in coopera-tion with the Indonesian authorities — remains indispensable for targeted prosecution of the most relevant actors.

---

[53] Personal interview with plantation company representative, Singapore, 28 February 2014.

[54] Centre for International Forestry Research. (2013). Q&A on fires and haze in Southeast Asia, Centre for International Forestry Research. Retrieved http://blog.cifor.org/17591/qa-on-fires-and-haze-in-southeast-asia#.U3QlDq32NMu, accessed 1 June 2016.

[55] Ouya, D. (2013). Less haze in Singapore as the cause becomes clearer and more complex, World Agroforestry Centre. Retrieved http://www.eurekalert.org/pub_releases/2013-07/wac-lhi070813.php, accessed 1 June 2016.

Conclusion

Now that the Transboundary Haze Pollution Act is in force, it seems inevitable that there will be a prosecution the next time a serious haze hits Singapore. This is simply because the Singapore public will not accept anything less. For a start, that entity to be charged is likely to be a Singapore-based oil palm operation that is either listed on the local stock exchange or, even if not publicly listed, is operated and run out of Singapore (hence, having a clear link to Singapore).

Of course, a successful prosecution could go a long way in mollifying an angry public, but there is no guarantee that it will resolve the problem. This is simply because there are too many other entities or plantation interests using fires to clear land in various parts of the sprawling Indonesian archipelago. Ultimately, strong Indonesian enforcement and prosecution action on the ground will matter most to resolve the problem at its core, together with cooperation mechanisms among regional states (including financial and technical assistance). Hence, the true value of the unilateral action by Singapore lies more in bringing pressure on Indonesia to take greater action. In other words, it takes the wind out of the sails of Indonesian officialdom's argument that the victim states should look to their own entities first.[56] In a nutshell, adversarial action before the Singapore courts is not a long-term solution. Any such expectation by the Singapore public would be naïve.

At the same time, it is conceivable that the tough sanctions provided by the Act might drive agri-business concerns out of Singapore altogether, particularly those that view themselves to be responsible and thus victimized. Singapore would thus have lost precious leverage over these actors. Overall, there could be a problem with action being taken against the wrong defendant or accused party or those that matter less in resolving the problem. Indeed, the publicly listed companies could de-list and easily relocate to places like Hong Kong and Malaysia, and still carry on with the *status quo* on the ground. It is not as if they will take the haze problem with them.

---

[56] Chua and Nazeer, see note 39.

Post-script

On 11 May 2016, Singapore's National Environment Agency (NEA) revealed that it had obtained a court order against a director of an Indonesian company suspected of engaging in fires and causing haze pollution.[57] The director was initially served with a legal notice to attend an interview with the NEA when he was in Singapore, but subsequently left Singapore without turning up for the interview. The court order thus allows the authorities to detain the director if he enters Singapore again.

The move is a further step under the Transboundary Haze Pollution Act to investigate individuals and companies involved in using fires. Six Indonesian-based firms have to date been given notices requiring them to provide information on the measures they are taking to combat and prevent fires on their land.[58] Of these, only two have replied. The NEA did not reveal the identity of the director and the company, although it has said that he is from one of the other four firms that have not replied.

However, Singapore's move to summon the director has not gone down well in Indonesia. Official quarters in Indonesia are reportedly taking issue with Singapore's extra-territorial application of its law in summoning the director.[59] In apparent retaliation, the Indonesian Minister for the Environment and Forestry was quoted as saying that Indonesia would henceforth seek to review existing and planned bilateral collaborations with Singapore, including on haze- and forest fire-related issues.[60]

---

[57] Straits Times. (2016, May 11). NEA obtains court warrant against director of Indonesian company with suspected haze links. *Straits Times*. Retrieved <http://www.straitstimes.com/singapore/environment/nea-obtains-court-warrant-against-director-of-indonesian-company-with, last accessed 1 June 2016.

[58] See note 20, above.

[59] Straits Times. (2016, May 13). Jakarta objects to Singapore moves against haze-linked firms. *Straits Times*. Retrieved http://www.straitstimes.com/asia/se-asia/jakarta-objects-to-singapore-moves-against-haze-linked-firms, last accessed 1 June 2016.

[60] Straits Times. (2016, May 15). Indonesia reviewing haze-linked collaborations with Singapore, says minister: Report. *Straits Times*. Retrieved http://www.straitstimes.com/asia/se-asia/indonesia-reviewing-haze-linked-collaborations-with-singapore-says-minister-report, last accessed 1 June 2016.

The Indonesian displeasure appears to be directed at the fact that Singapore was seeking to apply an extra-territorial law to net one of its nationals. One Indonesian government official even likened Singapore's act to harassing "our people".[61] While the identity of the individual and his firm has not been revealed, it may be assumed that the director is Indonesian, and that Indonesian officials are unhappy that Singapore appears to be targeting their countrymen instead of Singapore's own nationals or companies who may be involved in setting fires in Indonesia. Indeed, the six companies that have been given notices are Indonesian entities, as opposed to Singapore subsidiaries of Indonesian concerns (even though some of these Indonesian entities like PT Bumi Mekar Hijau are owned by parent conglomerates that have separate subsidiaries in Singapore).

That said, the Act makes no distinction between these different types of corporate holdings. In theory, the Act — being extra-territorial in nature — can be used against any foreign entity engaging in activities that cause haze pollution in Singapore, even if it had no operating base in Singapore. The six companies that were served notices were clearly those against which the Singapore authorities had the most compelling evidence.

It may also be that the company in question was able to use political levers to get its government to impose pressure on Singapore. Whatever the reason for the Indonesian ire, it seems clear that prosecution under the Act may be politically contentious, particularly if it targets well-connected Indonesian nationals. In addition, it is clear that cooperation between Singapore and Indonesia to target the relevant perpetrators of fires and haze cannot be assumed. Without such cooperation, the future prognosis for stamping out the recurring fires and haze problem is an alarmingly bleak one.

---

[61] Straits Times. (2016, May 24). Review of Singapore-Indonesia haze-linked programmes to be done by next week: Jakarta environmental official. *Straits Times*. Retrieved http://www.straitstimes.com/asia/se-asia/review-of-singapore-indonesia-haze-linked-programmes-to-be-done-by-next-week-jakarta, last accessed 1 June 2016.

21 The 2015 Game Changer

Audrey Tan

Journalist, The Straits Times, Singapore Press Holdings
audreyt@sph.com.sg

Tropical Singapore does not enjoy a seasonal climate, although it does suffer an unfortunate 'haze season' almost every year. As with many other countries in the region, Singapore has for decades been affected by smoke haze from forest fires in Indonesia, usually between the traditionally hot and dry months of June and October.

The haze-belching fires in Indonesia can be found on the concessions of oil palm or pulp and paper companies, although there has been debate about where the fires originate.[1] But whether the forests and carbon-rich peatlands are set alight by illegal land clearing methods, arsonists, wildfire, or stray cigarette butts, they all result in raging fires that have for years released tonnes of greenhouse gas emissions into the atmosphere,[2] and caused an unforgiving haze that affects Indonesia and its neighbors.

But it was not until 2015, when the haze crisis was the worst on record, that the region saw unprecedented levels of commitment to tackle the issue. In a sign of the severity of the crisis that year, all primary and secondary schools in Singapore were closed, for the first time due to the pollution, on September 25 2015. Across the region — especially at

---

[1] Cheam, J. (2015, September 15). New strategy needed to end haze problem. *The Straits Times*. Retrieved from http://www.straitstimes.com/opinion/new-strategy-needed-to-end-haze-problem

[2] Harris, N., Minnemeyer, S., Stolle, F., & Payne, O. (2015, October 16). Indonesia's fire outbreaks producing more daily emissions than entire US economy. Retrieved from http://www.wri.org/blog/2015/10/indonesia%E2%80%99s-fire-outbreaks-producing-more-daily-emissions-entire-us-economy

Ground Zero in Indonesia — the intense haze also caused illness and death, grounded flights, and affected tourism.

The cause of the exceptional haze? The El Niño weather phenomenon, which is linked to prolonged hot and dry weather in this region. It caused the forest and peatland fires in Indonesia to burn harder and for longer, resulting in an extended period of intense haze that hung over the region.

Tragic as this may be, the crisis also renewed determination on multiple fronts to stop the fires. Timing may have played an important role too. In October 2015, just two months before world leaders convened in Paris to ink a historic climate deal, the environment ministers of the ASEAN (Association of Southeast Asian Nations) countries set a target for the region to be haze-free by 2020, and agreed to develop an action-oriented, time-bound roadmap to achieve this.[3]

The governments of the region have moved together to act on the haze before, such as in 2002, when the ASEAN Agreement on Transboundary Haze Pollution was adopted. What was unprecedented last year however, was the extent to which individual governments — especially Indonesia — acted on the problem. The way companies and consumers both sought to become greener in their practices and purchases was also extraordinary.

Governments

During the height of the haze season in 2015, Indonesia president Joko Widodo cut short his visit to the United States to attend to the haze crisis back home.[4] He also told the British Broadcasting Corporation News (BBC) that the haze problem would be solved in three years, saying that Indonesia had already "gone to great lengths" to tackle this by sending soldiers, police officers and water-bombing planes to affected sites.[5] Another Indonesian

[3] Media Release: 11th Meeting of the Conference of the Parties to the ASEAN Agreement on Transboundary Haze Pollution. (2015, November 2). Retrieved from http://haze.asean.org/2015/11/media-release-11th-meeting-of-the-conference-of-the-parties-to-the-asean-agreement-on-transboundary-haze-pollution/

[4] Jokowi to stay in haze-affected regions. (2015, October 27). *The Jakarta Post*. Retrieved from http://www.thejakartapost.com/news/2015/10/27/jokowi-stay-haze-affected-regions.html

[5] Indonesia 'needs time' to tackle haze — Joko Widodo. (2015, September 29). *BBC*. Retrieved from http://www.bbc.com/news/world-asia-34387979

official assured Singaporeans in early 2016 that there will be "zero chance" that haze as severe as that in 2015 would affect the region. Nazir Foead, chief of Indonesia's newly formed Peatland Restoration Agency, said that the Indonesian government and other stakeholders have "full determination" to tackle the haze issue. Indeed, Indonesia never appeared to show more will in tackling the haze issue than it did in the wake of the 2015 haze.

The Peatland Restoration Agency, set up in January 2016 to restore about 2 million hectares of peatland in seven provinces by 2020, is one of the ways it is going about this. Widodo also proposed preventing new land from being used for oil palm plantations, urging producers instead to concentrate on using better seeds to increase their yields. The Indonesian police also arrested suspects and started investigations into agroforestry firms believed to have used illegal slash-and-burn land clearing methods. These efforts, observers say, are by far the most extensive taken by any Indonesian government to tackle the problem.[6]

This is true, but Widodo's commitment may all come to naught if his sentiments are not echoed by other members of the Cabinet. Indonesian Vice-President Jusuf Kalla, for example, told Indonesia's neighbors repeatedly to stop complaining about the haze, asking them to instead be grateful for the "nice air" they enjoy the rest of the year.

Such views, which appears to be shared by Indonesia's Environment and Forestry Minister Siti Nurbaya Bakar,[7] boil down to Indonesia's fear that action taken by other countries is an infringement of its sovereignty, especially when it comes to the forest fires, which are essentially a domestic problem. Neither has the government been transparent when it comes to land use information and those suspected of culpability. Indonesia's sprawling landscape is a minefield of overlapping land claims. In fact, different levels of its government — national, provincial and district — each have maps showing conflicting data. A centralized, public and accurate map is needed to help in the prosecution of errant companies. But

---

[6] Chan, F. (2015, October 3). Haze crisis set to be 'one of the worst on record'. *The Straits Times*. Retrieved from http://www.straitstimes.com/asia/haze-crisis-set-to-be-one-of-the-worst-on-record

[7] Arshad, A. (2016, June 14). S'pore can't take legal action against Indonesians over haze: Jakarta. *The Straits Times*. Retrieved from http://www.straitstimes.com/asia/se-asia/spore-cant-take-legal-action-against-indonesians-over-haze-jakarta

Indonesia still has not done this, although it has said it would. It has also not provided information on companies it is investigating, despite repeated calls from Singapore to do so.

This lack of transparency on both fronts has put the brakes on Singapore's anti-haze law, which was wielded for the first time in the wake of the 2015 haze. The Transboundary Haze Pollution Act was passed in Parliament in Singapore in 2014, and aims to punish those responsible for causing or condoning fires if burning results in unhealthy levels of haze in Singapore. Those found guilty under the Act can be fined up to 100,000 Singapore dollars a day, capped at a total of 2 million Singapore dollars, for causing unhealthy haze, defined as a 24-hour Pollutant Standards Index value of 101 or greater for 24 hours or more. At least seven companies have served noticed under this Act.

Under the Act, Singapore's National Environment Agency also obtained a court warrant against the director of an Indonesian company with suspected haze links, to secure his attendance when he enters Singapore. It was a move that Jakarta objected and reacted negatively to, saying it will review haze-related cooperations with Singapore.[8] Analysts have said that the only effective sanctions against errant companies are those that affect their finances.[9] But Indonesia's unwillingness to cooperate on this front has stalled progress to take action against companies who may have started fires or let their concessions burn.

Consumer action

Even as Singapore and other countries in the region prod Indonesia into doing more to address the problem, Indonesia has said that some companies are foreign-owned, and that its neighbors benefit from cheap agroforestry products like palm oil.[10]

[8] Arshad, A. (2016, May 16). Indonesia reviewing collaborations with Singapore over haze. *The Straits Times*. Retrieved from http://www.straitstimes.com/asia/se-asia/indonesia-reviewing-collaborations-with-singapore-over-haze

[9] Tan, A. (2016, May 9). Can S-E Asia be haze-free by 2020? *The Straits Times*. Retrieved from http://www.straitstimes.com/singapore/environment/can-s-e-asia-be-haze-free-by-2020

[10] Indonesia 'needs time' to tackle haze — Joko Widodo. (2015, September 29). *BBC*. Retrieved from http://www.bbc.com/news/world-asia-34387979

The blame has landed squarely on consumers, and their penchant for cheap products. In 2015, Singapore consumers responded by saying they would boycott products from haze-linked firms. Non-government groups such as the Singapore Environment Council and PM.Haze (People's Movement to Stop Haze) helped in this campaign, and supermarket chains also responded by yanking Asia Pulp and Paper (APP) — one of the firms believed to be linked to the haze — off their shelves.

It remains to be seen if Singapore consumers will continue buying green when the skies are blue and if haze-free products cost more, but the government has announced that it will take the lead by buying only printing paper products that bear the Singapore Green Label — which recognizes suppliers that practice sustainable forestry management, from September 2016.[11] However, this is just for pulp and paper products. Compared to the 4.5 million hectares of land taken up by pulpwood plantations in Indonesia, oil palm plantations take up 11 million hectares of land, according to a report by the US Department of Agriculture.[12] In this aspect, PM.Haze is intensifying its outreach efforts to drum up awareness on palm oil.

Still, the boycott and increasing consumer awareness had an impact on the agroforestry industry. Giants such as APP, and its rival Asia Pacific Resources International Holdings, have announced investments into fire suppression, canal damming, restoring peatlands and community programs to help villagers living near or within their concessions find alternative livelihoods.

These steps are laudable, but the fire-prone landscape does not respect concession boundaries — a fact that the confusion over the source of fires has highlighted. A recent *Straits Times* article put the percentage of Indonesia's forested peatlands that have been developed, deforested,

---

[11] Khew, C. (2016, April 14). SEC, Case applaud Government decision to buy only green label paper products. *The Straits Times*. Retrieved from http://www.straitstimes.com/singapore/environment/sec-case-applaud-government-decision-to-buy-only-green-label-paper-products

[12] Fogarty, D. (2015, November 4). Fix Indonesia's land use crisis to tackle the haze. *The Straits Times*. Retrieved from http://www.straitstimes.com/opinion/fix-indonesias-land-use-crisis-to-tackle-the-haze

drained, or burnt, as being more than half.[13] It would take concerted effort from the entire industry to work with each other and commit to more stringent standards. Without this, these measures would simply be seen as publicity stunts.

Conclusion

The 2015 haze is a game changer, as the unprecedented actions taken by governments, the industry, and consumers in the wake of the crisis have shown. This awakening is all well and good, but whether the haze problem can be effectively solved depends on how well action on the various fronts are sustained, and whether different countries and sections of society can work together.

The haze is a complex problem, and the fact that it has returned decade after decade points to the extent of the problem. This issue cannot be tackled unilaterally, although the Singapore and Indonesian governments can lead the effort with its policies, and for Indonesia, by being transparent with information. Without synergy, action taken on various fronts would simply go up in smoke.

[13] Fogarty, D. (2015, November 4). Fix Indonesia's land use crisis to tackle the haze. *The Straits Times*. Retrieved from http://www.straitstimes.com/opinion/fix-indonesias-land-use-crisis-to-tackle-the-haze

22 Battling the Haze of Cross-Boundary (Mis)Governance in Transboundary Air Pollution: A Perspective from Singapore

Eugene K. B. Tan

Associate Professor of Law, School of Law,
Singapore Management University
eugene@smu.edu.sg [1]

The haze, which has regularly blanketed Singapore and parts of Malaysia and Indonesia in the past two decades, is a poignant reminder that while innovative laws and nimble diplomacy are necessary, they are insufficient in dealing with the annual scourge of transboundary air pollution from Indonesia.

To be sure, tough measures are needed. But political will within the polluting state(s), states being polluted, consumer awareness and responsibility, and the corporate will to move decisively away from business-as usual are as crucial in getting to the root of this longstanding problem. Singapore's Foreign Minister Vivian Balakrishnan read the riot act to stakeholders on the persisting haze problem:

> Let me be frank, and perhaps undiplomatic. I know foreign ministers are supposed to be nice, smiley, diplomatic people. But since I used to

---

[1] The author participated in the legislative debate on the Transboundary Haze Pollution Bill on 5 August 2014 in his capacity as a Nominated Member of Parliament in Singapore's 12th Parliament.

be the Environment Minister, I can be frank. Brutally frank. The transboundary haze that has afflicted our region for far too long is a man-made tragedy and a crime. A man-made tragedy and a crime. It would be bad enough if this was a natural disaster. We would all shake our heads, huddle together, reach out, help one another. But this is not a natural disaster. This is a deliberate, man-made tragedy. Vandalism against society, against the environment, and ultimately, against ourselves. It has impaired the health of millions of people, compromised the safety of aircraft, and damaged our regional economy. This year, it has been estimated that more than 2 million hectares of forests, a lot of which was growing on peatland — and you know that peatland represents millennia-worth of carbon captured in our ground — more than 2 million hectares of forests have been set alight. Huge quantities of CO_2 have been released, estimated at around 1.6 billion tons of CO_2. You know, it pains me to have to travel to Paris to negotiate global agreements, whilst right here in our backyard, we are releasing huge amounts of global greenhouse gases. It puts things in context. And, all this is happening because actually only a handful of people, a handful of big companies, are really profiting from this entire exercise at the expense of the environment and of the rest of society. This is a classic example of privatising the gain and socialising the pain.[2]

This reflection essay is organized as follows. The first part discusses Singapore's Transboundary Haze Pollution Act, a law enacted in 2014. The next part considers the role of supply chain management as well as consumer awareness and action. The role of cooperation at the regional level, especially through the auspices of Association of Southeast Asian Nations (ASEAN) is examined in the third part. The fourth part concludes. The essay argues that the haze problem requires political will at the individual state level but also at the regional level. A multi-stakeholder approach is needed to resolutely tackle the longstanding problem, which can also deal with the governance deficits (or misgovernance) at various parts of the supply chain of palm oil and paper and pulp.

---

[2] Transcript of Minister for Foreign Affairs Dr Vivian Balakrishnan's plenary address at the 4th Responsible Business Forum on Sustainable Development, 3 November 2015; available at https://www.mfa.gov.sg/content/mfa/overseasmission/geneva/press_statements_speeches/2015/201511/press_20151103.html (accessed on 5 June 2016).

Going after the perpetrators: Singapore's Transboundary Haze Pollution Act

In 2014, Singapore's Parliament enacted the Transboundary Haze Pollution Act ('the Act') to enhance our ability to deal with transboundary haze.[3] The law covers both acts of commission and omission, whether they occur within or outside of Singapore, which resulted in haze pollution in Singapore.

So far, there have been no prosecutions under the law although there have been enforcement actions that are still in progress.[4] It should be no surprise if public expectations in Singapore are high that this law is a game-changer and that the problem of the haze will be resolved before long. But the realities are more complex, and that is where pathos and even despair may quickly be the dominant feeling. If only such a challenging international problem like the haze can be simply solved by domestic legislative fiat!

Given that the serious episodes of air pollution in 1994, 1997, 2006, 2010, 2013, and 2015 are transboundary in nature, this law would not be of any use if it did not provide for extra-territorial application (as though the proscribed act was committed in Singapore). Yet, extra-territoriality, which enables Singapore authorities to exercise power beyond Singapore's territorial limits, is double-edged because it is often regarded by another state as an infringement of its national sovereignty.

The dynamics of Indonesia–Singapore bilateral relations complicate the picture further. At one level, the lack of meaningful cooperation on the haze may be, at another level, be a manifestation that Indonesia objects to this law having such extensive extra-territorial reach.[5] This is notwithstanding that the law does not provide Singaporean courts with the jurisdiction to determine if a foreign state is responsible for transboundary haze pollution in Singapore, an internationally wrongful act.

---

[3] Act 24 of 2014. The Act was passed by Parliament on 5 August 2014 and assented to by President Tony Tan Keng Yam on 10 September 2014. The law was brought into force on 25 September 2014.

[4] At the time of writing: May 2016.

[5] At another level, the lack of cooperation on the haze issue points to the sensitivities that continue to affect bilateral ties between Indonesia and Singapore.

Neither does it seek to impose liability on a foreign state for the damage caused to Singapore and Singaporeans as a result of the haze pollution.

The Act does not just cover landowners whose lands are the sources of transboundary haze pollution but also extends to other actors, including those engaged to start the fires and those involved in the management of an offending entity. It does not seek to pinpoint a sole perpetrator, recognizing that transboundary haze pollution is very much the culmination of a series of deliberate acts in a decision making process that is cross-border and involving different parties.

It is highly significant that the law not only creates an entirely new offence but also broadly expands the notion of extra-territorial reach. It achieves this by criminalizing a first entity's failure to prevent transboundary haze caused by a second entity to which the first entity participates in the management of the second. This attempt to deal with the chain causation is a significant extension of jurisdiction well beyond that what is found in our general criminal law and existing legislation with extra-territorial reach.[6]

Notwithstanding the law's extra-territorial reach, the law is limited to entities with a presence in Singapore. This ability to investigate and prosecute companies carrying on part of a business here in Singapore, irrespective of where they are registered, is important. This levels the playing field by requiring foreign companies to adhere to the same standards of care and responsibility in preventing transboundary haze pollution as Singapore-registered and Singapore-based companies.

However, the biggest challenge to the law's successful implementation is that Indonesia's strong cooperation is necessary for any successful prosecution and enforcement. For instance, most of the evidence needed to mount a prosecution is likely to be found in Indonesia, where the companies or individuals allegedly conducted the illegal activities that directly caused the haze. Such evidence includes geospatial information such as the ownership and occupation of the land. This is notwithstanding the law providing for the power to obtain information, the power to examine and to secure attendance of those who may know of matters related to an alleged offence.

I am skeptical that there will be a successful prosecution under the law. Land use and land tenure in Indonesia is governed by a complex

---

[6] A good example is the Prevention of Corruption Act (Cap. 241, 1993 Revised Edition).

web of national, provincial and customary laws that often compete and even conflict with each other. As such, it is a legal nightmare to pinpoint with a high degree of certainty as to who might be responsible for the air pollution.

Furthermore, as fires move across the landscape, propelled by topography and wind, one cannot assume that landowners are burning within their concessions where the hotspots are identified. Overlaying concession maps with hotspot locations may be insufficient to meet the burden of proof required for a successful prosecution in our courts.

Whether out of national pride, or denial of the scale of the problem, or both, Indonesia has been reluctant to take up the Singapore government's consistent offer of additional anti-haze assistance since 2005.[7] What all this means is that the ground realities are a lot more complex and the haze problem all the more challenging in light of Indonesian intransigence. This means that the law's actual impact is likely to be limited, more apparent than real.

Singapore recognizes the complexities and constraints in Indonesia. It has been careful not to point fingers at the Indonesian authorities but instead seek to assist Indonesia within her means while being sensitive to the dynamics of Indonesian domestic politics and bilateral relations of both countries.

Supply chain management and consumer awareness and activism

Beyond law enforcement, it is timely to determine how the law can be enhanced, especially in encouraging companies with direct and indirect interests and/or stakes in the plantation business (especially paper and pulp and palm oil). This requires such companies to be scrupulously responsible for their

---

[7] The latest rebuff, at the time of writing, was in June 2016. Indonesian Vice-President Jusuf Kalla said that environmental issues needed to be dealt with through a regional agreement, not a bilateral one. In May 2016, Indonesian media reported that Indonesia would scrap ongoing and upcoming collaboration with Singapore on environment, forestry, and haze-related issues as part of a larger unilateral review of bilateral cooperation with Singapore. See "Indonesia rebuffs Singapore offer of haze assistance," *Today* (Singapore), 10 June 2016, p. 6.

supply chains by constantly reviewing their supply chains, including demanding fully traceable raw materials in their supply chains. Given the gravity of the haze problem, it cannot be business as usual: A race to the bottom.

Such businesses should strongly consider the implementation of effective anti-air pollution programs by evaluating the sustainability of their supply chains, including detecting and responding to improper environmental conduct. Implementing and maintaining a sustainable and responsible agribusiness remains a prudent and recommended course of action. This is not just to avoid the long arm of the law and negative publicity but because it is also the right thing to do.

In any case, consumers are becoming savvy and discerning in their purchasing choices. If a business is not able to show responsible and sustainable sourcing in their supply chain, then their sales and reputations could be detrimentally affected. There was a flurry of activity domestically following the 2015 edition of the haze (such as the non-sale of Asia Pulp and Paper (APP) paper products in Singapore, the Green Label scheme being reviewed) but the moment the winds stopped blowing the haze in our direction, we took our feet off the pedal. This may well reflect a stoic resignation that we are at the mercy of our neighbors. However, there is much that active and enlightened consumerism can do in self-help to not only mitigate the problem but to push the corporate entities to act responsibly.

There is a lot that Singapore can do to tackle the recurring haze problem, especially with regards to consumer education, our procurement policies in the public and private sectors, as well as ensuring that Singapore and Singapore-based financial institutions are diligently practicing responsible financing practices. We have made significant inroads with the Transboundary Haze Pollution Act and will introduce a sustainable procurement policy for the public sector.

But we are not moving fast enough to generate the momentum to demonstrate unequivocally that we have a right to clean air and environment. There is a lot more that can be done in terms of scope and speed. Although we must not lose sight of the structural constraints in Indonesia as well as the internal political dynamics, it is also equally important to focus on what is within Singapore's control, especially taking on the corporate transgressors.

We have tended to respond in a knee-jerk, piecemeal manner when comprehensive, resolute and sustained action involving multiple stakeholders are urgently needed to tackle the haze scourge. Here the government has to take the lead, and it is about involving all stakeholders to hit the errant companies where it hurts most — their financial bottom-line. For example, banks in Singapore have until 2017 to put in place robust governance systems under the Association of Banks in Singapore (ABS) Guidelines on Responsible Financing introduced in October 2015.[8] This is way too long. The ABS guidelines provide a bare-bones framework for banks in determining, assessing and managing environmental and social risks in their provision of banking and financial services. The guidelines are not prescriptive enough to make banks to be found culpable more easily than before. They are too generic and lightweight and only require banks to demonstrate that they have adopted the three stipulated principles on responsible financing in their business models. There does not seem to be any indication that external audits would be conducted. Not withstanding the ABS guidelines, the litmus test is whether banks are committed, in form and in substance, to a responsible financing policy and their compliance?

It is a baby step taken by ABS at a time when bolder measures are needed to deal with the scourge that haze is. For instance, 12–18 months is too long a time frame for banks to publish their environmental, social and corporate governance (ESG) policy framework! There is no need to reinvent the wheel. There are readily available tools like the well-established Equator Principles, which banks can easily adopt.[9] But the more pertinent question is to what standard banks, of their own accord, are prepared to hold themselves to. Are banks prepared to subject themselves to the rigorous process when determining, assessing and managing environmental and social risk in the provision of banking and financial services? Otherwise, banks will engage in responsible financing only in form but not in substance.

---

[8]The Guidelines are available at http://www.abs.org.sg/docs/library/abs-guidelines-responsible-financing.pdf (accessed on 5 June 2016).

[9]On the Equator Principles, see http://www.equator-principles.com (accessed on 5 June 2016).

It would be a sad and bitter irony if banks with Singapore banking licenses are bankers to the very companies causing or contributing to the haze pollution here!

Can banks be found to be contravening the Singapore's Transboundary Haze Pollution Act, if they are lending to companies that are found to have slash-and-burn activities contributing to the haze?

A bank is unlikely to fall within the ambit of the Act merely for lending to companies that are found to have engaged in activities contributing to the haze. This is a weakness of the Act that can be easily rectified.

However, under the Act, the bank could be liable if it is proven to have participated in the management of a company that was engaging in conduct that caused or contributed to haze pollution in Singapore.

Alternatively, it may be possible to take legal action under civil law against a bank that provided banking and financial services to a company for the purpose of its activities that caused or contributed to haze pollution in Singapore. The bank could be said to have not exercised a duty of care, especially if the ABS Guidelines and/or the bank's own policies are not complied with.

Perhaps the most useful deterrent is fear that the banks' reputation and business will be severely undermined if they are found to be complicit in the activities of their clients that caused or contributed to haze pollution in Singapore. This is where consumer awareness and activism has a vital role to play.

In similar vein, the Singapore government must unequivocally demonstrate its resolve in tackling the haze problem by ensuring that those who benefit from a business presence in Singapore do not undermine our society's wellbeing and economic vitality. The government must take a strong lead by adopting a robust procurement policy that only procures goods and services from sustainable and responsible sources.

Singaporeans would also expect no less from our government-linked companies, including Temasek Holdings (Private) Limited and GIC Private Limited (previously known as the Government of Singapore Investment

Corporation Private Limited), to not be directly or indirectly responsible for the haze. Can more be done to ensure that their investments and business partners in the plantation supply chain, are not engaging in conduct that is detrimental to the health of Singaporeans and our economy and to Southeast Asia at large? This highlights the centrality of supply chain management in the fight against transboundary pollution.[10]

In this connection, all stakeholders should adhere to the precautionary approach. The precautionary principle stipulates that where threats of serious or irreversible damage to the environment or human health have been identified, the lack of full scientific certainty is not a reason to postpone cost-effective measures that can help prevent or reduce environmental degradation or damage to human health.

Cooperation in ASEAN

Bilateral cooperation and multilateral cooperation within the ASEAN framework can and must play a far bigger and more important role in tackling resolutely transboundary haze pollution, a longstanding problem in Southeast Asia that shows no signs of abating. In particular, observing the letter and spirit of the 2002 ASEAN Agreement on Transboundary Haze Pollution will be a good start. All 10 ASEAN member states are signatories to this important treaty.[11] Furthermore, the goal is for ASEAN to be haze-free by 2020, a goal set by the member states environment ministers. It will take a monumental effort for ASEAN to be haze-free by 2020.

The haze problem is a complex challenge, and time is of the essence in preventing further deterioration of the problem. Much is at stake. Given the gravity of the clear and present threat and the fact that the problem shows little improvement all these years, they point to the work that is cut

---

[10] See Temasek Holdings' letter to Singapore's Chinese daily *Lianhe Zaobao*, "Temasek takes concrete action for the well-being of our communities," 28 September 2015. This letter was in response to my remarks in "Temasek could affect business practices through its investment and divestment decisions," published in "Academics: Government procurement can lead the way in solving haze problem," *Lianhe Zaobao*, 27 September 2015. Temasek's letter is available online at http://www.temasek.com.sg/mediacentre/medialetters?detailid=23683
[11] Indonesia ratified the ASEAN Agreement on Transboundary Haze Pollution in early 2015. However, it did not seem to matter when it came to the crunch.

out for ASEAN while also suggesting that all is not well. What helps, however, is that an ASEAN-wide framework is in place.

What is needed is to muster the political will to make actions count. Otherwise, the regional agreement and haze control mechanism are merely a paper tiger. To get the target back on track, there must be a meeting of the minds among ASEAN leaders and people that the haze is a critical issue that deserves the full attention of all ASEAN member states and that each and every one of us can make a difference.

The ASEAN haze mechanism looks promising on paper but it is also, so far, distinguished by its singular ineffectiveness. There is much to be done at the national and regional level but, unfortunately, the issue has not gained the undivided attention it deserves when most ASEAN countries are caught up with other domestic issues.

To be fair, Indonesia has sought to respond to their neighbors' concerns. In 2015–2016, several policies and initiatives were launched, including the establishment of the Peatland Restoration Agency and a new moratorium on granting palm oil concessions. These are to be welcomed although questions are legitimately being raised about these policies and initiatives are more form than substance. Are they more appeasement and symptomatic treatment of a systemic issue? Are the Indonesians committed to the long haul? That remains to be seen.

I certainly hope they will be implemented with the requisite haste, effectiveness, and efficacy. However, successful implementation can be challenging in the Indonesian context due to the decentralized governance structure, the deep vested interests in maintaining the status quo, endemic corruption, and it is not at all clear that there is the political will to make a decisive move away from business and governance as usual. It is a case of two steps forward and one step backward.[12]

---

[12] See, for example, Natasha Hamilton-Hart, "Multilevel (mis)governance of palm oil production," *Australian Journal of International Affairs*, 69(2) (2015): 164–184. Hamilton-Hart argues that the governance failures associated with the palm oil industry stem from different stakeholders' competing interests in contexts of highly unequal wealth and power distribution. Misgovernance is not an unintended consequence of institutions failing to keep up with markets in scale and scope, but is embedded in the multilevel governance regime that supports, and partially regulates, the industry.

In lieu of a conclusion: Beyond pathos and despair

We have to manage our expectations and not expect overnight dramatic changes with regards to the haze problem. If Indonesia is willing to seek foreign technical assistance, that can help scale up the means of and resolve in addressing this transboundary challenge. We have a right to breathe air that does not sicken or kill us, or harm our economic livelihood. Indeed, Article 28(f) of the ASEAN Human Rights Declaration provides for "the right to a safe, clean and sustainable environment." In similar vein, the ASEAN Charter recognizes sustainable development, as one of the purposes of ASEAN is to "ensure the protection of the region's environment and the sustainability of its natural resources".[13] A multi-stakeholder collaborative approach is needed if the transboundary haze is to be consigned to the dustbin of history. Anything less will likely mean that the scourge will remain with us for some time to come. The issue has the makings of not just an environmental disaster, but also a human security threat. The damage will not only be to environment, economy and people's health but also to the reputation and standing of ASEAN and the individual member states.

---

[13] Article 1(9), ASEAN Charter.

In lieu of a conclusion: beyond pathos and despair

23 Recent Forest Fires and the Effect of Indonesian Haze in the Philippines

Maria Luisa G. Valera

Assistant Professor, Department of Economics,
College of Economics and Management,
University of the Philippines Los Baños
mgvalera@up.edu.ph

The highlights of this paper are two fold: (1) a brief report of the recent concluded forest fires from 2014 to first quarter of 2016 and (2) summary of the effect of Indonesian haze in the Philippines.

Recent forest fires from 2014 to first quarter of 2016

In the Philippines, about 8.04 million hectares or 27 percent of the total land area are covered with forest (World Development Indicator, 2015) — roughly 9 percent higher compared to 15 years ago. This can be attributed to the reforestation efforts by the government.

Looking back, in the late 1990s forest fires occurred mostly in northern and central parts of the country. The month of March is the peak of the fire season and is considered the hottest temperature that causes the drying up of grassland. And majority of the incidents were attributed to human factors such as negligence and carelessness.

However, in recent reports from 2014 to first quarter of 2016, the northeastern parts of the country had the highest incidence of forest fires. The months of March and April are still the peak of the fire season but surprisingly there were three incidents of forest fires that occurred in August 2014. These were caused both by nature and man-made activities.

Table 1–3 summarizes the incidence of forest fires for years 2014 to first quarter of 2016 using historical method excerpts from different news articles. The reports contain the date, place, description, affected area and cause of forest fires incidence in the Philippines.

Table 1.   Incidence of Forest Fires (2014) in the Philippines

Date	17–25 February 2014
Place	Eastern Samar: Giporlos, Balangiga, Quinapondan, Salcedo, Lawaan, Marabut, Basey, Guiuan (*eastern part of the country*)
Description	Located in the Eastern Visayas region; with total land area of 466,047 hectares; mostly hit by typhoons because it faces the Philippine Sea of the Pacific Ocean
Affected	More than 53 hectares of forest lands, non-residential zero
Cause	Super typhoon Yolanda's damage to plants and trees that withered during hot season
Action	Aerial spray to prevent the fire from spreading
Date	19–21 March 2014
Place	Mt. Banahaw, Mt. Cristobal (*northeastern part of the country*)
Description	Mt. Banahaw is an active volcano while Mt. Cristobal is a dormant volcano in Luzon protected landscape covering 10,901 hectares of lands and they exist side by side
Affected	At least 50 hectares of cogon in Mt. Banahaw; 100–140 hectares of grasslands in Mt. San Cristobal; endangered plants and animals
Cause	Intruders (backpackers) into the restricted area (preservation or reforestation area); lack of forest rangers; fire fighters had no capability to fight due to the magnitude of the area
Action	Died on its own
Date	3–6 August 2014
Place	Rapu-Rapu, Albay (Poblacion, Morocborocan, Sitio Acal Mananao, Mananao, Guadalupe, Buenavista, Sitio Minto, San Ramon, Batan Island (*northeastern part of the country*)
Description	Approximately with 5,000 households; town is rich in mineral deposits, such as copper, gold, zinc and coal; Rapu-Rapu and Batan lie to the east of Luzon together with the islands of San Miguel and Cagraray (that form the northern rim of Albay Gulf)

(*Continued*)

Table 1.   (*Continued*)

Affected	Almost 2,093 hectares of grassland; 1,748 families were evacuated; mat-weaving livelihood of residents; water system
Cause	Kaingin (slash-and-burn farming); aggravated by the dried cogongrass and trees after the devastation brought by Typhoon Glenda
Action	Activated the Municipal Disaster Risk Reduction and Management Council (MDRRMC) in order to intensify monitoring and preparedness and to prepare the evacuation sites; aircraft equipped with a carrier filled with 33,000 liters of water dropped in the burning area
Date	9–10 August 2014
Place	Mt. Isarog, Naga City (*northeastern part of the country*)
Description	Mt. Isarog has a rich diversity. It displays four major types of natural habitat or vegetation; from the warm grassland and lowland forest to the wet and cool climate of montane forest
Affected	At least 150 hectares of second-growth forest: 100 hectares in the village of Kaway-nan in Tinambac town and 50 hectares in the village of Harubay; hundreds of young trees of hardwood species
Cause	Undetermined origin
Action	Quick response of provincial disaster risk reduction and management office of Camarines Sur and volunteers from their village: quickly studied the direction of the wind and fire and preempted its spread by just clearing areas that were vulnerable

Table 2.   Incidence of Forest Fires (2015) in the Philippines

Date	19–20 March 2015
Place	Mount Santo Tomas watershed (portion); Sitio Bigis, Kabuyao village in Benguet's Tuba town (*northern part of the country*)
Description	Tourist spot; inactive volcano
Affected	Three hectares of forest and grasslands in an area below two giant satellite dishes
Cause	Undetermined origin
Action	Philippine Air Force helicopter dropped water on the areas

(*Continued*)

Table 2. (*Continued*)

Date	April 16–18, 2015
Place	Mountain village of Mapita in Aguilar: three days; Natividad town, Pangasinan (Caraballo mountains): 18 August (16 hours) (*northern part of the country*)
Description	Tourist spot
Affected	At least 300 hectares of forest near the mountain of Mapita in Aguilar
Cause	Kaingin (slash-and-burn farming); inadequate firefighting equipment; inaccessible area (too high); lack of personnel to supervise; very low budget for forest protection
Action	Watch it from the lowlands and get ready; did nothing on top

Table 3. Incidence of Forest Fires (2016) in the Philippines

Date	26–31 March 2016
Place	Mt. Apo in Kidapawan City, North Cotobato and Digos City, Davao del Sur (*southern part of the country*)
Description	The phases of Mt. Apo in Kidapawan City and Digos City are adjacent to each other; with an area of 54,974 hectares of land; was declared a natural park in 2004
Affected	About 300 hectares of grassland and forested areas; massive movement of wild animals (boars, rats, lizards, birds and deer)
Cause	Might be campers who had set up a camp fire; dry weather and wind
Action	(a) Regulated the number of climbers and prohibited them to use firecrackers, burn debris and set up campfires; woods, logs and charcoals are not allowed for cooking; (b) recommended the suspension of the Mt. Apo summer trek during summer; (c) created firebreaks when a brushfire begins (firebreaks: unbroken lanes from six to ten feet wide, cleared of all vegetation)
Date	29 March–1 April 2016
Place	Mt. Kanlaon in La Castelana, Negro Occidental (near the crater) (*southern part of the country*)
Description	Tourist spot; it is one of the active volcanoes in the Philippines and part of the Pacific Ring of Fire
Affected	About 400 hectares of grassland
Cause	Ashed releases and fiery rocks and other incandescent materials (volcano's intermittent minor eruptions)
Action	69 firefighters

The El Niño phenomenon together with illegal entries of campers were blamed for the forest fires that happened in Mt. Banahaw and Mt. Cristobal on March 2014, with protected landscape (preservation or reforestation area) covering more than 10,000 hectares in Luzon. This was aggravated by the dried debris after the devastation brought about by Typhoon Glenda (international code: Rammasun) in mid-July 2014. Moreover, nature struck again in Mt. Kanlaon in Negros Occidental by end of March 2016 due to the intermittent minor eruptions that release ashes, fiery rocks and other incandescent materials.

The cause of forest fires by man-made activities was not only for monetary purposes like the common slash-and-burn farming known as "kaingin" that was blamed during the forest fires in Rapu-Rapu, Albay on August 2014; but also by the lack of discipline and the curiosity of tourists or campers. For instance, the forest fires in Mt. Banahaw and Mt. Cristobal mentioned earlier were caused by illegal entries of tourists or campers who had set up a campfire. This negligence by campers once again caused the forest fires in Mt. Apo on March 2016. Efforts had been made to regulate the number of campers and restrict the entries in reforested area but admittedly there were not enough forest rangers to patrol. In addition to that, fire fighters had no capability to fight due to the magnitude of the area and the needed equipment was not available.

From 2014 to early 2016, more than 5,400 hectares were damaged due to forest fires. More than 4,400 hectares were damaged in 2014, about 300 hectares in 2015, and roughly 700 hectares in the early months of 2016. Total amount of costs was not reported but it affected the ecosystem of the endangered plants and animals, the growing of hundreds of young trees of hardwood species, as well as the livelihood of residents who were forced to evacuate their residences during the event.

A number of actions must be taken seriously by the local and national government starting from instilling in the minds and hearts of the village folks the act of volunteerism, discipline, and loyalty to protect their own forestlands against intruders, campers, and also against their self-vested interests. Quick response of the disaster risk reduction and management office of the province with ample fire fighters or forest rangers equipped with the right equipment that can go beyond the magnitude of area. For instance, the availability of the standby aircraft equipped with carriers filled with liters of water in case the need arises. But this is subject to budgetary concern that usually hinders such prevention of forest fires to spread and stop at once. For instance, about 660 billion Philippine pesos or 14.3 million US dollars is required for firefighting equipment, resources and professionals to combat a 3,000-hectare of forest fires (Hoon *et al.*, 2015). Lobbying for budgetary alignment is deemed necessary in the deliberation of local or provincial budget in order for the reforestation and prevention of forest fires to be given much attention. But it requires a lot of support and effort from all concerned parties to act as one and be united.

The effect of Indonesian haze in the Philippines

The haze from Indonesian forest fires greatly affects its neighboring countries, Malaysia and Singapore. The southern and central parts of the Philippines have been mildly affected as well because of their proximity to Indonesia.

Mindanao, southern part of the Philippines, is more than 1,200 kilometers or 745 miles away from the nearest Indonesian forest fires. Specific areas in the country that had been affected include Cebu, Cotabato, Davao, Leyte, General Santos, Cagayan de Oro, Palawan, and Negros Occidental. It affected the country for almost a month from 4 October 2015 to 28 October 2015.

The direction of the wind causes such transboundary haze from Indonesia going to the Philippine territory. Specifically, monsoon winds blowing northeast from the fire incident carried the smog toward the central part while southwest monsoon winds to the southern part of the country. The Typhoon Lando (international code: Koppu) that hit the northern part also contributed to the entry of haze.

The effect of haze had been mild compared to other parts of the Southeast Asia. One of its direct effects was on air traffic due to very poor visibility that led to the cancellation of flights in Mindanao (e.g., Philippine Airlines and Cebu Pacific flights from Cotabato to Manila were cancelled). It also alarmed the public about the increased risk of respiratory infections and cardiac ailments. Health offices required those highly exposed to haze to wear facemask (especially for those with asthma).

The unity not only in Association of Southeast Asian Nations (ASEAN) countries but worldwide cooperation in the reforestation effort would surely pave the way in preventing forest fires from occurring. Strict regulations and compliance by all, from individual households to the regional level are necessary in the success of saving the forest.

References

Aguirre, E., Dinoy, O. and Magbanua, W. (2016, March 28). Mt. Apo fire contained in North Cotabato side, rages on in Davao Sur side. *Philippine Daily Inquirer*. Retrieved http://newsinfo.inquirer.net/776563/mt-apo-fire-contained-in-north-cotabato-side-rages-on-in-davao-sur-side

Aguirre, E., Dinoy, O. and Magbanua, W. (2016, March 29). Camping climbers blamed for Apo fire. *Philippine Daily Inquirer*. Retrieved http://newsinfo. inquirer.net/776588/camping-climbers-blamed-for-apo-fire

Aguirre, E. and Magbanua, W. (2016, March 27). Mt. Apo forest fire still raging as of Sunday night, threatens more areas. *Philippine Daily Inquirer*. Retrieved http://newsinfo.inquirer.net/776380/mt-apo-forest-fire-still-raging-as-of-sunday-night-threatens-more-areas

Anda, R. D. (2015, October 27). Rains, lush forest blunt effects of haze in Palawan. *Philippine Daily Inquirer*. Retrieved http://newsinfo.inquirer. net/734655/rains-lush-forest-blunt-effects-of-haze-in-palawan

Cardinoza, G. (2015, April 21). Execs say firemen helpless vs. forest fires. *Philippine Daily Inquirer*. Retrieved http://newsinfo.inquirer.net/686701/execs-say-firemen-helpless-vs-forest-fires

De Jesus, J. L. (2014, February 14). Forest fires scorch Eastern Samar. *Philippine Daily Inquirer*. Retrieved from http://newsinfo.inquirer.net/579664/forest-fires-scorch-eastern-samar

De Jesus, J. L. (2014, February 25). Forest fires in Eastern Samar put out — report. *Philippine Daily Inquirer*. Retrieved http://newsinfo.inquirer.net/580524/forest-fires-in-eastern-samar-put-out-report

Escandor, J. Jr. (2014, August 9). Fire destroys 30 hectares of forest in Camarines Sur. *Philippine Daily Inquirer*. Retrieved http://newsinfo.inquirer.net/627708/fire-destroys-30-hectares-of-forest-in-camarines-sur

Escandor, J. Jr. (2014, August 10). 150 hectares of forest consumed by fire on Mt. Isarog. *Philippine Daily Inquirer*. Retrieved http://newsinfo.inquirer.net/627817/150-hectares-of-forest-consumed-by-fire-on-mt-isarog

Escandor, J. Jr. (2014, August 10). Volunteers quick to act vs fire on Mt. Isarog. *Philippine Daily Inquirer*. Retrieved http://newsinfo.inquirer.net/627869/volunteers-quick-to-act-vs-fire-on-mt-isarog

Fernandez, E. O. (2015, October 22). Haze forces cancellation of flights in Mindanao. *Philippine Daily Inquirer*. Retrieved http://newsinfo.inquirer.net/733769/haze-forces-cancellation-of-flights-in-mindanao

Fernandez, E. O. (2015, October 23). Haze cancels flights. *Philippine Daily Inquirer*. Retrieved http://newsinfo.inquirer.net/733795/haze-cancels-flights

Fernandez, E. O. (2016, February 15). DENR to ARMM farmers: Stop 'kaingin' and drought. *Philippine Daily Inquirer*. Retrieved http://newsinfo.inquirer.net/764797/denr-to-armm-farmers-stop-kaingin-amid-drought

Fernandez, E., et al. (2016, March 29). Strong winds rekindle fire in Mt. Apo; wild animals flee. *Philippine Daily Inquirer*. Retrieved http://newsinfo.inquirer.net/776953/strong-winds-rekindle-fire-in-mt-apo-wild-animals-flee

France-Presse, A. (2015, October 4). Philippines suspects week-long haze from Indonesia fires. *Philippine Daily Inquirer*. Retrieved http://newsinfo.inquirer.net/727818/philippines-suspects-week-long-haze-from-indonesia-fires

Gomez, C. P. (2016, April 1). Kanlaon fire under control. *Philippine Daily Inquirer*. Retrieved http://newsinfo.inquirer.net/777285/kanlaon-fire-under-control

Gomez, C. P. and Udtohan, L. (2015, October 26). Wear face masks, Palace tells people exposed to haze. *Philippine Daily Inquirer*. Retrieved http://newsinfo.inquirer.net/734425/wear-face-masks-palace-tells-people-exposed-to-haze

Gonzales, Y. V. (2015, October 28). Ph is now haze-free, says DOST-Pagasa. *Philippine Daily Inquirer*. Retrieved http://newsinfo.inquirer.net/735103/ph-is-now-haze-free-says-dost-pagasa

Hoon, A. W., Loo, J. and Cherian, J. (2015, October 26). Transboundary haze: Asean needs to act. *Philippine Daily Inquirer*. Retrieved http://business.inquirer.net/201303/transboundary-haze-asean-needs-to-act

Magbanua, W. (2016, March 27). Massive forest fire rages on Mt. Apo; hikers flee inferno. *Philippine Daily Inquirer*. Retrieved http://newsinfo.inquirer.net/776347/massive-forest-fire-rages-on-mt-apo-hikers-flee

Mallari, D. T. Jr. (2014, March 21). Forest fires damages mystic Mt Banahaw. *Philippine Daily Inquirer*. Retrieved http://newsinfo.inquirer.net/587672/forest-fire-damages-mystic-mt-banahaw

Mangosing, F. (2015, October 26). 'Light haze' in Metro Manila possibly from Indonesia fires, says Pagasa. *Philippine Daily Inquirer*. Retrieved http://globalnation.inquirer.net/129975/light-haze-in-metro-manila-possibly-from-indonesia-fires-says-pagasa

Manlupig, K. (2015, October 19). Thick haze blankets skies in Davao City, CDO, GenSan. *Philippine Daily Inquirer*. Retrieved http://newsinfo.inquirer.net/732614/thick-haze-blankets-skies-in-davao-city-cdo-gensan

Mier, M. A. (2014, August 3). Bush fire destroys 6,000 hectares of forest, grasslands. *Philippine Daily Inquirer*. Retrieved http://newsinfo.inquirer.net/ 625962/bush-fire-destroys-6000-hectares-of-forest-grasslands

Mier, M. A. (2014, August 5). Forest fire still rages in Albay town. *Philippine Daily Inquirer*. Retrieved http://newsinfo.inquirer.net/626311/forest-fire-still-rages-in-albay-town

Mier, M. A. (2014, August 6). Salceda: 'God took care' of Albay forest fire. *Philippine Daily Inquirer*. Retrieved http://newsinfo.inquirer.net/626659/ salceda-god-took-care-of-albay-forest-fire

Mier, M. A. (2014, August 7). Forest fires break out anew in Albay. *Philippine Daily Inquirer*. Retrieved http://newsinfo.inquirer.net/627158/forest-fires-break-out-anew-in-albay

Mier, M. A. (2014, August 9). Bush, forest fires continue to rage, threaten 3 Albay towns. *Philippine Daily Inquirer*. Retrieved http://newsinfo.inquirer. net/627610/bush-forest-fires-continue-to-rage-threaten-3-albay-towns

Nawal, A. (2016, March 29). Mt. Apo fire spreads to 300 hectares of forest, grassland. *Philippine Daily Inquirer*. Retrieved http://newsinfo.inquirer.net/776731/ mt-apo-fire-spreads-to-300-hectares-of-forest-grassland

Pazzibugan, D. (2015, October 28). 'Smaze' starting to clear — Pagasa. *Philippine Daily Inquirer*. Retrieved http://newsinfo.inquirer.net/734918/ smaze-starting-to-clear-pagasa

Quismundo, T. (2015, November 20). Indonesian VP: Excuse our haze, blame the wind. *Philippine Daily Inquirer*. Retrieved http://globalnation.inquirer. net/132373/indonesian-vp-excuse-our-haze-blame-the-wind

Quitasol, K. (2015, March 20). Forest fire hits portions of TV soap's Benguet location. *Philippine Daily Inquirer*. Retrieved http://newsinfo.inquirer. net/680195/forest-fire-hits-portions-of-tv-soaps-benguet-location

Sabillo, K. A. (2015, October 27). Aquino: PH won't castigate Indonesia for haze, to offer help instead. *Philippine Daily Inquirer*. Retrieved http://globalnation. inquirer.net/130011/aquino-ph-wont-castigate-indonesia-for-haze-to-offer-help-instead

Soeriaatmadia, W. (2015, October 20). Haze crisis could persist into new year, say experts. *The Straits Times*. Retrieved http://www.straitstimes.com/asia/se-asia/ haze-crisis-could-persist-into-new-year-say-experts

World Development Indicator. (2015). Forest area (% of land area). Retrieved http://data.worldbank.org/indicator/AG.LND.FRST.ZS/countries

Yan, G. (2016, March 31). How firebreaks, educating the public can save PH mountains. *Philippine Daily Inquirer*. Retrieved http://newsinfo.inquirer. net/777234/how-firebreaks-education-can-save-ph-mountaints

24 Singapore's Transboundary Pollution Act 2014: Prospects and Challenges

Helena Varkkey

Senior Lecturer, Department of International and Strategic Studies,
University of Malaya
helenav@um.edu.my

June 2013 saw Singapore battling with its most severe episode of haze yet. During that period, Singapore's Pollutant Standards Index hit an all-time record high of 401. This event ignited a diplomatic row between Indonesia and Singapore, with Singapore's then Minister of the Environment and Water Resources Vivian Balakrishnan almost immediately accusing Indonesia of not caring about the welfare of its neighbors (Grant & Bland, 2014). It was also around this time that Balakrishnan first revealed plans to table a Transboundary Haze Pollution Act that would provide for criminal and civil liability for any Singaporean or non-Singaporean entity causing or contributing to transboundary haze pollution in Singapore (Woo, 2014a).

The first act of its kind

The Act was formally announced by the Minister on 19 February 2014, and opened to public consultation till 19 March 2014 (Woo, 2014a). The Act then was reviewed and tabled in Parliament, ultimately being passed by Parliament on 5 August 2014 and assented to by the President on 10 September 2014. It has now formally come into operation since 25 September 2014 (Government Gazette, 2014).

The Act is unique for its application of extra-territoriality; it covers the operations of all Singapore and non-Singapore entities whose activities outside of Singapore contribute to haze pollution in the city-state. The Act is the first of its kind for Singapore, as Singapore usually only punishes action overseas only for severe crimes, such as corrupt acts or illegal sex with minors. It is also the first of its kind in the region and the world, as there is currently no law in the world that allows a country to prosecute commercial entities in other countries for such offences (Woo, 2014a). Currently, the only way to catch entities based overseas is if somebody in the entity's management position comes to Singapore (Mediacorp 5, 2014).

The Act makes it a criminal offence when an entity engages in conduct, or authorizes any conduct which causes or contributes to haze in Singapore. Penalties ranging from 50,000 Singapore dollars to 100,000 Singapore dollars, up to a maximum aggregate fine of 2 million Singapore dollars can be imposed if the entity has deliberately ignored requests by authorities to take appropriate action to prevent, reduce, or control the pollution (Government Gazette, 2014). An individual company officer can also be held personally responsible.

Affected parties may also bring civil suits against errant entities. The civil damages recoverable under the Act will be determined by the courts of Singapore based on personal injury, physical damage, or economic loss (Straits Times, 2014a). Civil action can also be taken against errant entities by industries (such as aviation, tourism and construction) if they can prove that they have suffered serious economic consequences (Kotwani, 2014).

An important inclusion in this Act are its Presumptions, included in Part II Section 8 (Government Gazette, 2014). Presumptions allow the court to assume that a fact is correct until it is proven otherwise. Since proving what happens abroad is difficult, evidential Presumptions relating to causation (linking open burning elsewhere and wind direction with the presence of haze in Singapore) and culpability (based on ownership and occupation of land) have been inserted (Kotwani, 2014), and help to give teeth to this Act. Among others, it importantly allows for reliance on satellite imagery, meteorological information, and maps as evidence (Tay & Chua, 2014). The Presumptions place the burden of proof on entities to provide a rebuttal through their own land maps. Hence, a company can defend itself by proving the fires were caused by natural disasters or by parties not under its direction. Companies can also rebut the Presumptions by proving that the concession maps used by the authorities are wrong (Tay & Chua, 2014).

Positive prospects

With companies having to prove that they are not liable for the haze, they should feel more pressure to be transparent and responsible. Hence, this Act should incentivize companies to be more forthcoming about their landholding and practices, especially with regards to sharing their internal land maps and concession maps with the authorities. This has been a long-standing problem between the firms and the authorities, and this shift in the burden of proof could potentially resolve this issue once and for all.

The Act would also surpass the diplomatic need of going through government channels when faced with wrongdoing. Since the Act allows individual lawsuits directly against companies, it will be able to bypass any 'friction' that might occur if the matter has to go through governments (interview with Tay, S. in Woo, 2014b). As governments have usually resorted to diplomatic consultations with home countries when foreign companies were suspected of burning, most of these cases were resolved diplomatically and not legally. The Act will also be able to address criticisms that Singapore is lenient towards suspected companies that are headquartered in Singapore but operate in Indonesia. It could also stop Indonesians from using the excuse that Singapore-linked companies get away with environmental destruction in Indonesia (interview with Syarif, L. M. in Straits Times, 2014b).

There should also be positive developments in related industries, like banking and investment, which come with this Act. With the Act in place, banks and investors should become more careful when approving loans to companies since these financial institutions would not want to expose themselves to more risks of civil or criminal liability (interview with Tay, S. in Woo, 2014b). Hence, this Act should encourage more financial institutions to evaluate loans based on sustainability and reputational risk.

Weaknesses and challenges

Despite the allowances for Presumptions made in the Act, identifying errant companies may still be problematic. Firstly, there is no incentives or protection offered for whistle-blowers under the Act (interview with Ang, P. H. in Straits Times, 2014b). And the extra-territorial nature of the

Act means that the Singaporean National Environment Agency (NEA) would need to work closely with their counterparts in Indonesia to build a case against these companies. There is concern that the authorities may not be able to work well together (Tan & Bassano, 2014).

Indonesia has thus far shown a weak track record; clearing land through burning is prohibited in Indonesia but authorities have so far only successfully prosecuted a handful of companies for starting such fires. Enforcement of land clearing laws is weak in Indonesia due to lack of skilled ground staff to assess remote areas. Furthermore, corruption is rife in Indonesia, and companies have been able to continue burning large tracts of forests every year, clearing it for planting trees for palm oil or paper production. Indonesia has also denied Singapore's request to make digitized land-use maps and concession maps of fire-prone areas publicly available under ASEAN Haze Management System (HMS), citing privacy and legal limitations. Availability of maps are thus currently under an ad-hoc basis (National Environment Agency, 2014), which means that Singapore will have to individually request for maps of areas as and when fires are suspected, generally slowing down and further complicating the process of assigning blame.

Once these companies are brought to court, building a case against these companies would be an equally challenging proposition. In terms of criminal liability, proving causation or contribution of an entity's conduct to haze pollution in Singapore would be problematic. The prosecution would have to prove beyond reasonable doubt that the haze was at least partially from that particular fire. Factors like the thickness of the smoke, trajectory, and its ability to travel far distances all come into play and will have to be evidenced in court (interview with Chun, J. in Business Times, 2014).

In terms of civil cases, a claimant would have to show probable cause that his personal injury, disease, mental or physical incapacity or death is due to the conduct of the defendant, which may also be challenging. Firstly, there may be a time lag between the haze and the personal injury. Secondly, haze may not be the only contributing factor to that injury. Also, the defendant's conduct may have caused just a few fires compared to the total number of hotspots at the time. The court may find it difficult to attribute the exact proportion of the defendant's responsibility to the person's injury (interview with Chun, J. in Business Times, 2014).

Of course, culprits may exploit any loophole to absolve themselves of blame. For example, entities might claim that they had no control over sub-contractors who start the fires (interview with Ong, B. in Straits Times, 2014b). This is a common reason given by companies when confronted by authorities. Furthermore, the question remains if the current fines are high enough to truly deter would-be culprits, especially compared to the profits they would earn otherwise[1] (Chua, 2014).

Test case: The 2015 haze

The severe haze episode in 2015 was a chance to see the newly minted Transboundary Haze Pollution Act in action. The first step as provided for in Part III Section 9 of the Act is to send out Preventive Measures Notices to suspect companies, requesting them to undertake preventive measures, discontinue or not start burning activities, and submit a plan of action to extinguish or prevent the spread of fire (Government Gazette, 2014).

The NEA sent out six such notices in September and October 2015 (Channel NewsAsia, 2015b). It also requested information directly from the Indonesian government for more names of companies suspected of contributing to haze pollution (Channel NewsAsia, 2015a). While Indonesia has yet to formally respond, Singapore has since received responses from two of these companies, and is in the midst of reviewing the information provided before deciding if charges can be pressed (Channel NewsAsia, 2015b). It is interesting to note that one of the companies singled out under this new Singaporean Act, Asia Pulp and Paper, recently had one of its suppliers, PT Bumi Mekar Hijau, cleared in Indonesian courts of causing fires in Sumatra, a decision that was met by shock and disappointment in Indonesia as well as worldwide (Ismail, 2015). Such is the potential value of this new Act, if a highly suspect case is cleared in Indonesia, as is often the case, there is still a chance that the Singaporean Transboundary Haze Pollution Act can still enforce justice.

---

[1]Other non-monetary punishments which may have been more impactful and damaging, especially in terms of company image would have been a demotion in the status of companies, compulsory reforestation projects, seizing any assets the entities may have in Singapore, and even prohibitions from doing business in Singapore.

However, as of this writing, there have been no further developments or indications that the Act has been enforced beyond the delivery of Preventive Measures Notices and requests for information. As an Act that has newly come into operation, there will no doubt be teething issues as prosecutors figure out the best way to collect evidence and press charges.

Outlook for the future

This Act is certainly an interesting test case to explore the effective reach of extra-territoriality in transboundary pollution issues. However, as discussed above, the effectiveness of this Act will depend much on the cooperation extended from Indonesian agencies. One avenue in overcoming this challenge is if the Singaporean authorities are able to work together and share information with non-governmental organizations (NGOs) like Greenpeace, who have compiled considerable evidence from their own investigations (interview with Ang, P. H. in Straits Times, 2014b).

Despite this slow start, it is hoped that this Act will send a strong signal of deterrence to potentially errant companies, especially considering the public relations storm that could be unleashed upon companies that are exposed of such irresponsible conduct. Furthermore, the Act indirectly puts pressure on Indonesia to step up its own efforts to resolve the haze problem. As a whole, the Act shows that Singapore is willing to take action where it can. Public views have been generally supportive, and environmentalists and observers have lauded the Act as a positive step forward in tacking the haze menace. Even the neighboring Malaysia has expressed interest in developing its own similar extra-territorial act to address transboundary haze pollution (Palansamy, 2015). The people of Southeast Asia and the world are no doubt waiting with bated breath to see if this Act can finally deliver the punch that the region needs to conclusively tackle transboundary haze pollution.

References

Business Times. (2014, February 20). Govt proposes law to fight transboundary haze pollution. *Business Times*.

Channel NewsAsia. (2015a, October 7). Indonesia accepts Singapore's offer of haze assistance package. *Channel NewsAsia*.

Channel NewsAsia. (2015b, October 12). NEA sends notice to 6th Indonesian firm over haze. *Channel NewsAsia*.

Chua, G. (2014, August 4). Parliament: Transboundary Haze Bill penalties too small, say MPs. *Straits Times*.

Government Gazette. (2014). *Transboundary Haze Pollution Act 2014*, Republic of Singapore: Singapore.

Grant, J. and Bland, B. (2014, February 19). Singapore widens battle against toxic haze from forest fires. *Financial Times*.

Ismail, S. (2015, December 30). PT Bumi Mekar Hijau found not guilty by Indonesian court of causing forest fires in Sumatra. *Channel NewsAsia*.

Kotwani, M. (2014, February 28). Experts say draft haze Bill won't solve issue in long term. *Channel NewsAsia*.

Mediacorp 5. (2014, February 19). Public feedback sought on Bill targeting firms that cause haze. *Mediacorp*.

National Environment Agency. (2014). Factsheet on Transboundary Haze Pollution. Singapore. http://www.nea.gov.sg/docs/default-source/corporate/COS-2014/transboundary-haze-pollution.pdf (accessed on 5 June 2017).

Palansamy, Y. (2015, October 21). Putrajaya looking to adopt Singaporean law on transboundary haze. *Malay Mail*.

Straits Times. (2014a, February 19). Govt proposes new haze law aimed at errant parties; public views sought. *Straits Times*.

Straits Times. (2014b, February 21). S'pore, Indonesia experts say 'yes' to law on transboundary haze. *Straits Times*.

Tan, D. and Bassano, M. (2014). Dissecting the Transboundary Haze Pollution Bill of Singapore. Retrieved from Columbia.

Tay, S., and Chua, C. W. (2014, February 21). To end the haze problem, both penalties and cooperation are needed. *TODAY*.

Woo, S. B. (2014a, February 19). Haze: Govt seeking views on new bill to fine local, foreign companies responsible. *TODAY*.

Woo, S. B. (2014b, February 20). Government proposes law to allow action against firms causing haze. *TODAY*.

Editor's Contributions

25 Tackling Haze with Cost Sharing*

Parkash Chander and Euston Quah

http://www.straitstimes.com/singapore/tackling-haze-with-cost-sharing
Published
Jun 13, 2014, 11:03 PM SGT
Reproduced from The Straits Times (Opinion)

The seasonal haze in South-east Asia, caused by fires to clear land in Indonesia, has affected air quality in neighbouring Singapore and Malaysia for years. Indeed, it has become an almost annual occurrence.

Severe haze is expected again this year because of the likely drought caused by the cyclical El Nino weather pattern.

In fact, the Singapore Government has reportedly stocked a huge number of face masks as a precautionary measure.

Tradable pollution rights?

Since the haze is a case of transboundary pollution, most of the conventional tools for controlling pollution cannot be applied. The key sticking point is the sovereignty and independence of both polluted and polluting countries.

Professor Roland Coase, in an important paper that helped him win the Nobel Memorial Prize in Economic Sciences in 1991, argues that assignment of tradable pollution rights to either the polluter or the polluted can lead to optimal control of pollution.

He asserts that optimal control of pollution can be achieved irrespective of whether the polluter has the right to pollute or the polluted has the

---

*The contents in this chapter were originally published in *The Straits Times* written by Parkash Chandar and Euston Quah.

right to clean air. Since the right to pollute is a property right that has value, if the right is tradable, the result will be optimal control of pollution at least cost to society.

In the case of South-east Asian haze, however, the Coasian solution cannot be applied as there is no supranational authority to assign and enforce pollution rights. A polluting country can be pressured but not forced to reduce its pollution.

Therefore, only voluntary negotiations among the affected countries can solve the problem.

Since Singapore and Malaysia cannot enforce their right to clean air, the South-east Asian haze is a case in which the polluting country, Indonesia, has the right to pollute. A 2002 agreement to get Asean countries to implement measures to prevent forest fires has failed. Of the 10 member states, Indonesia is the only country which has yet to ratify the Asean Agreement on Transboundary Haze Pollution.

Prerequisites for success

To be successful, a regional agreement must take into account several factors.

First, no country should be worse off after the agreement is implemented.

This can be ensured if the cost of controlling pollution is smaller than the total damage that the affected countries suffer from pollution.

The cost of controlling pollution can then be distributed in such a way that each country is better off. By this we mean that each country's share of the cost is smaller than the damage it would have had to suffer if pollution were not controlled.

Estimates of the cost of the haze to South-east Asia vary, but a conservative estimate made by the Asian Development Bank (ADB) was US$9 billion for the 1997 episode.

No similar estimates for the cost of controlling and preventing fires that lead to haze are available.

But a rough estimate can be obtained by taking into account the average number of hectares that are cleared by fire each year and the average cost of clearing a hectare by non-burning methods. An educated guess for these costs is approximately US$1.2 billion (S$1.5 billion).

The second factor to be considered when formulating a regionwide agreement is that no group of countries should be better off by leaving the agreement. Clearly, the two non-polluting countries — Singapore and Malaysia — would only be better off by leaving the agreement if the amount they pay to control the haze is higher than the damage it causes.

The agreement should also be such that neither Singapore and Indonesia, nor Malaysia and Indonesia, can be made better off by leaving the multilateral agreement and establishing a separate bilateral accord.

The third prerequisite of a successful agreement is that it should not be possible to draw up an alternative agreement such that no country is worse off, but some country or countries are better off.

Game theory

There is a branch of game theory concerned precisely with the problem of designing such agreements. It has been shown that if the cost of controlling pollution is shared by the affected countries in proportion to the damage they experience from pollution, the three above-mentioned conditions for a successful agreement can be satisfied.

This cost-sharing rule has come to be known as the Chander-Tulkens rule.

An earlier study by Professor Euston Quah and others, published in the 2004 issue of Britain-based academic journal *Environment And Planning*, has estimated the relative impact of the haze on Indonesia, Malaysia, and Singapore to be 93.8 per cent, 5.1 per cent and 1.1 per cent, respectively. Thus, the Chander-Tulkens rule requires Indonesia, Malaysia and Singapore to contribute to the cost of controlling the haze in the same proportions.

Since the estimated cost is US$1.2 billion, Indonesia, Malaysia, and Singapore should contribute approximately US$1.125 billion, US$61.2 million, and US$13.2 million a year, respectively.

The sum of US$1.125 billion a year may seem to be a huge sum for a developing country like Indonesia. But it is much smaller than the approximately US$8.447 billion damage a year that it will be able to avoid.

Implementation

There may, of course, be some practical difficulties. The countries will need to agree on how the US$1.2 billion should be deployed.

One approach could be to use the money to enhance the ability of the Indonesian authorities to detect, locate and respond to the fires, as well as strengthen its ability to prosecute those responsible.

Another approach could involve paying subsidies to encourage land clearing by non-burning methods.

Perhaps a combination of two of the approaches would work best.

Commonly heard suggestions to control fires and the haze, such as boycotting Indonesian products (products made from palm oil, for example), will not work. They simply penalise well-behaved plantations as well as errant ones.

Furthermore, palm oil products may also be used as intermediate goods in the production of final goods, in which case the approach also penalises other firms that are not complicit in the fires.

The recently suggested imposition of a strict liability law with regard to fires emanating from plantations is a step forward simply because it incentivises plantation owners to be aware of their actions.

But the strict liability law proposed by Singapore only applies to Singapore-owned plantations.

Besides causing haze, the fires are the largest single contributor to Indonesia's greenhouse gas emissions.

In 1997, the haze episode over the three-month period actually exceeded the annual carbon emissions of Europe.

Since Indonesia has committed to reduce its greenhouse gas emissions by 26 per cent (or 41 per cent with international assistance) by 2020, it would make perfect sense for Indonesia, Malaysia and Singapore to join forces to promote clearing of Indonesia's land by non-burning methods as a project under the clean development mechanism provided for in the Kyoto Protocol.

Any international assistance received for the project could be subtracted from the cost of controlling and preventing fires in Indonesia.

Only the net cost need be shared by the three affected countries in the proportions proposed above, thus benefiting them all.

stopinion@sph.com.sg

The first writer is a fellow of the Econometrics Society and professor at the Jindal School of Government and Public Policy in India. The second writer is professor and head of economics at Nanyang Technological University. He is also president of the Economic Society of Singapore.

26 Pollution Controls as Infrastructure Investment*

Euston Quah and Joergen Oerstroem Moeller

http://yaleglobal.yale.edu/content/pollution-controls-infrastructure-investment
Published
Thursday, October 6, 2016
Reproduced from YaleGlobal

SINGAPORE: Despite political pressures over many years and various enforcement measures, palm oil producers in Indonesia continue to slash and burn to clear land, harassing neighboring countries with transboundary pollution.

Simple economics may offer a new approach for slash-and-burn agriculture, which if successful might also have relevance for similar environmental encroachments. Farmers and plantations must find it profitable to ditch slash and burn, and those demanding a haze-free life must contribute financing. Benefits and costs must be designed in such a way that no alternative exists making countries better off — creating a win/win situation for everybody taking part.

Palm oil is the world's most competitive vegetable oil with global production more than doubling since 2000 and expected to grow even more as advanced economies favor natural oils over artificial trans-fats for health reasons. The palms, native to Africa, were transferred to Malaysia in the 20th century and later to Indonesia.

Today, that region produces more than 80 percent of the world's palm oil.

---

*The contents in this chapter were originally published in *The Straits Times* written by Euston Quah and Joergen Oerstroem Moeller.

Economic losses in health, tourism and canceled flights along with school and business closures are an indisputable consequence of haze, but do not hit those responsible. The Asian Development Bank estimated the 1997 costs of haze to be US$9 billion, and that does not include the negative images of deforestation and pollution, highlighted by activists and in turn leading to decreased foreign direct investment. Often, the burning of fields gets out of control, triggering wildfires with impacts that are difficult to estimate.

The cost of switching to non-burning methods is estimated at US$1.2 billion.

The only way to change producers' behavior is bridging the cost difference between slash-and-burn versus non-burning methods. Those who resent the pollution must invest $1.2 billion to avoid the $9 billion in costs.

Uniform cross-the-board measures won't work. About 40 percent of the haze originates from plantations run by large corporations and 60 percent from small landholders including subsistence and indigenous farmers. The large corporations have the financial resources to ditch slash-and-burn methods, but the small operators do not and cling to traditions.

Market forces won't work. Slash-and-burn is less costly than environmentally friendly methods relying on manpower, heavy machinery or new technologies. Under current market conditions — low prices due increasing supply amid falling demand as well as currency volatility and competition from soybean oil — the initial investment will, at best, be profitable only in the long run and burden the producer with a short-term cash drain.

If forced to switch to expensive methods and if the governments could enforce regulations, small operators would close. Lacking coordination or influence, small farmers cannot easily hike the market price for palm oil and they resist pressure, legislation and rules. Politicians are understandably reluctant to target a major contributor to the domestic economy, estimated at about 5 percent of Indonesia's GDP.

Governments and environmentalists could disrupt the economic calculus by paying rewards or subsidies to small landholders that use haze-free methods. Such a system anchored in financial rewards or subsidies, aiming to improve methods and increase enforcement, would change the economic calculus. High rewards for those who act swiftly could create a group of pioneers who demonstrate the advantages. New technologies and synergies might eventually reduce cost differences and the need for subsidies, but development of new technology is held back by a lack of economic incentives. Such a system would not avoid payouts to large corporate plantations, and organizers could borrow from methods in European countries that encourage collecting waste to generate energy.

The large corporations could do more to reduce haze, but many hide behind the small farmers and the government is lenient. Media reports this year disclosed fires in concession areas managed by six companies.

The only solution is to finance the costs for small stakeholders to switch methods of land clearance. It may seem unfair — like extortion — to ask the victims to pay out to avoid being harmed. But realistically, many in Southeast Asia may be willing to pay to end the annual choking haze that can last three to four months. Such a policy conforms with welfare economics: A policy enhances welfare if those who are better off can compensate those who lose and still be better off.

If such a policy were left to countries directly involved in financing, then the amount should be distributed according to relative negative impact. A 2004 study suggests that about 94 percent of the victim damages are in Indonesia, 5 percent in Malaysia and 1 percent in Singapore. Game theory economics suggests that if victims share the costs of controlling pollution in proportion to their damages then no country would be worse off, no country should be better off by leaving the agreement and no alternatives exist that make some countries better off.

But political realities rarely work as smoothly as economic theory. Haze has so far been classified as an environmental problem, relevant only for those producing the pollution and those suffering from the effects. Such a narrow focus might be replaced by classifying such haze as an element in economic and social development. As such, preventing the burning and the economic consequences belongs among the many activities undertaken by developed countries as well as international institutions like the World Bank, the Asian Development Bank and the newly born Asian Infrastructure Investment Bank under the label of development assistance.

Lending expertise and financial support would improve environmental conditions in Southeast Asia, saving the region's countries billions of dollars while modernizing production methods used primarily by small landholders. The benefits of transforming this agricultural sector in Malaysia and Indonesia might deliver spinoffs to other sectors in the economy including agricultural machinery. And over the long term, the health benefits might be considerable as is the case for switching to a more modern production structure and obliterating the image of a sector mired in outdated methods. Another substantial if nontangible benefit: Accusations by activists about the sector not being a good corporate citizen would no longer be warranted.

The world is trying to get a handle on climate change and environmental problems with limited success. Agreements to limit carbon dioxide emissions as a result of the Paris agreement, to go in effect in November, are encouraging but not enough.

A rewards system for small stakeholders in economies may serve as a template for similar cross-border environmental problems to analyze production methods, technology and market structure and develop financing and enforcement systems, making it profitable for polluters to switch into cleaner production methods.

The underlying assumptions include that people are willing to pay for not enduring pollution, small stakeholders will respond to economic incentives, and international financial institutions are prepared to classify reduction of pollution as infrastructure improvements and therefore eligible for support. If so, the world may have a much-needed tool to act against global pollution.

Euston Quah is professor and head of economics with Nanyang Technological University, Singapore. He is also president of the Economic Society of Singapore. Joergen 'Oerstroem Moeller is visiting senior fellow, ISEAS Yusof Ishak Institute, Singapore. He is also adjunct professor with the Singapore Management University and Copenhagen Business School and an honorary alumni of the University of Copenhagen.

27 What Can Singapore Do about the Haze?*

Euston Quah and Tan Tsiat Siong

http://www.businesstimes.com.sg.ezlibproxy1.ntu.edu.sg/opinion/
what-can-indonesia-do-about-the-haze
Published
Wednesday, October 28, 2015 — 05:50
Reproduced from Business Times (Opinion)

Over the past two months, the haze has generated a spate of heated responses in Singapore. As we now eagerly await the nearing end to the choking air, we are well aware that the problem is far from its resolution, and we would certainly expect the haze to revisit us in the subsequent years.

Instead of incessantly pointing our fingers at our neighbour, we should focus on how we could possibly escape this recurring conundrum.

It is clear that Singapore has little transnational jurisdiction on the haze culprits apart from Singapore firms, which might or might not be a sizeable proportion of the total number of errant plantations. Holding Singapore firms accountable is justified on both moral and legal grounds, but without disclosures of land ownership from Indonesia, whether this would reach fruition is very much debatable.

Besides, if Indonesia and other countries do not penalise their own culprit firms to an equivalent extent, Singapore firms will bear the burden unfairly in relative terms.

Secondly, a handful of firms in Singapore recently signed a declaration that their products are free of raw materials from suppliers being

---

*The contents in this chapter were originally published in *The Straits Times* written by Euston Quah and Tan Tsiat Siong.

investigated for the Indonesian fires. There have also been calls for consumers to boycott haze-linked products.

These are applaudable moves, but are unlikely to bring any measure of success in reducing the fires. This is primarily because Singapore's consumption of paper or palm oil products constitutes a small portion of global demand. Additionally, it is unlikely that all local firms and consumers will comply as they can free-ride on others; hence the effect is further reduced.

Unless we can get most local firms and consumers, as well as our neighbouring countries, to practise the boycott, the likely outcome would be higher prices for a seemingly futile attempt. Here is where behavioural economics can lend a hand to change consumers' behaviour. This also warrants a study on the effects of a boycott on the profits of the Indonesian plantations and the suppliers of products linked to these plantations.

Moreover, pressures levied against Indonesian goods might result in trade retaliation.

If these same boycotting measures were to be employed in the major consuming countries (namely China, India and even Indonesia), we could be more hopeful. Nevertheless, would China and India, who are not directly affected by the haze, boycott polluting products at their own expense? Would domestic consumers in Indonesia be willing to pay higher prices for palm oil and paper products in exchange for cleaner air? Such questions are important and require further study.

For all of the above, the facts must first be established. Who are the real perpetrators? The issue of boycotting products from plantations that are being investigated, but subsequently not proven guilty, also arises. As is often the case in law, the accused is not guilty until proven so.

In this regard, Mr Han Fook Kwang, in an Oct 18 commentary in *The Sunday Times*, suggested shifting the onus of evidence to the companies, meaning that the defendant is presumed to be guilty until proven otherwise. But as with the case before, it is very difficult to prove either way. This idea of the alleged party having to clear its own name may result in too many convicted cases. For small suppliers and plantation owners, it may prejudice their market survivability, paving the way for the market to be dominated by a few suppliers.

Thirdly, Singapore has provided Indonesia with fire-fighting assistance, but both the offer (by Singapore) and acceptance of help (by Indonesia) could have been much quicker.

Another preventive approach would be for Singapore and other victim countries to provide monetary side payments to Indonesia, on the condition that plantations adopt land-clearing practices without the use of fires. The amount of side payment will have to depend on the calculated costs of alternative land-clearing methods. The share in the cost by victim countries, including Indonesia, should be in proportion to the damage suffered.

Given these limited options, there is really not much Singapore can do to touch the root of the problem.

But what Singapore can really do in the meantime, while waiting for mitigation efforts to take place in Indonesia, is to adapt. This is similar to the climate change issue, where adaptation becomes more important than ongoing mitigation, which takes a long time to be realised.

This year, during the school closures in Singapore, some students reportedly still turned up for school. Some among them did not receive the notification; for others, their parents were unable to take time off work to look after them. In this aspect, there is definitely room for improvement in information dissemination, and in designing and encouraging work-from-home and study-from-home arrangements.

On the days where PSI readings have lingered in the "Very Unhealthy" and "Hazardous" ranges, people were spotted on the streets without masks, or wearing surgical masks that are ineffective against fine particles. This calls for improved ways to educate or nudge the general public, which could reduce the incidence of haze-related illnesses and the loss of productivity. The authorities could even impose a mandatory shutdown of outdoor sports facilities when needed, accompanied by a compensation scheme for affected businesses.

It is to be noted that while there has been no shortage of N95 masks, air filters in certain major electrical appliance stores are sold out. On hindsight, since we have expected the severe El Niño episode and the impending haze from international reports, a system of information dissemination could have been set up so that businesses and consumers can take the appropriate action before the arrival of the haze.

To understand the haze problem, we need contributions from multiple disciplines — the business, science and medical, political economy and legal aspects, including estimating the damage costs of the haze for all victim countries in the aggregate as well as sectorally, and in knowing how much people are willing to pay to avoid the polluted air.

Damage costs must necessarily be computed for categories such as loss of life, physiological and psychological health impact in the short and long term, loss of productivity, impact on sectors such as tourism, retail, food and beverages, aviation, conferences and conventions and its mutiplier effect, loss of intangibles (such as scenic views), threats of land subsidence from forest clearing, contribution to global warming, loss of goodwill and reputation.

Losses could also take the form of unwillingness of expats or foreign workers to work in Singapore, which may result in a tighter labour market, especially in the professional sectors. If the haze is a permanent recurring feature, it could also mean that more remuneration would have to be offered to attract future workers. All these may result in higher costs of business, as well as cost of living.

The key is that damage estimates must be continually updated and convincing to all affected countries.

We have seen good progress in Singapore's response to the haze this year, but as long as the problem persists, there is undeniably more to be done.

While there is a lot that Indonesia needs to do on its own, it is perhaps more timely that Singapore focuses on adaptation strategies. For now, it appears that many proposed solutions on the part of Singapore would likely be futile.

Professor Euston Quah is president of the Economic Society of Singapore and head of economics at Nanyang Technological University, where Tan Tsiat Siong is a PhD research student.

28 What Can Indonesia Do about the Haze?*

Euston Quah and Tan Tsiat Siong

http://www.businesstimes.com.sg.ezlibproxy1.ntu.edu.sg/opinion/what-can-indonesia-do-about-the-haze
Published
Thursday, October 29, 2015 — 05:50
Reproduced from Business Times (Opinion)

If there is any solution to the relentless haze problem, it lies in Indonesia.

President Joko Widodo's administration has undoubtedly shown renewed resolve and increased efforts in tackling the Indonesian forest fires, but his three-year target to greatly reduce the fire episodes seems more of a political timeframe than a realistic one.

We have no wish to be right about this, but there are good reasons to believe it is so.

The Indonesian authorities have begun arresting executives from firms accused of being responsible for the fires. The penalties for those found guilty are heavy, but convicting the culprits is easier said than done. The perpetrators first have to be accurately identified, which is already a daunting task, and they may include small farmers instead of just big corporations.

The lack of land ownership information adds to the challenge, not to mention the local and central government's vested interests in protecting the industries (worth an estimated US$18 billion); corruption and coordination and enforcement failures also figure in the mix.

In addition, the time required to secure a hearing in court and thereafter, convicting the culprits, will likely be long. Hard evidence proving the sources of fires, required for litigation, would be difficult to collect.

---

*The contents in this chapter were originally published in *The Straits Times* written by Euston Quah and Tan Tsiat Siong.

As it appears, the punishment is severe — but neither swift nor certain.

Drawing the line

Within Indonesia, which has a large consumer market (unlike Singapore), banning products from alleged culprit plantations would still be futile for the same reason — that it is tedious to track down culprits and accord responsibilities.

Moreover, palm oil and paper are intermediate goods, meaning they are inputs to the production of other goods and services. Should consumers then boycott other products that use materials from pollutive plantations, thus penalising firms not directly complicit with the act? Will this information even be available? Where should we draw the line down the supply chain?

We must also remember that the forest fires are both a man-made and nature-caused problem.

Since the intense El Niño had already been expected early on, Indonesia could perhaps have informed Asean of the impending haze problem, the steps it planned to undertake and raised a request for regional support.

On the local front, more resources and logistical support could have been diverted to empower the local communities for proactive fire prevention, or at least for a faster reactive response to fires.

Small fires are easier to contain. Responses should be prompt and coordinated with standard operating procedures, focused objectives, and non-overlapping responsibilities across institutions.

President Joko's idea of creating water canals might also work in reducing the dryness of peatland, and thus the chances of nature-caused fires, but this will not work where peatlands have been intentionally drained to make way for plantations.

However, with most of these fires being man-made, solutions can largely be successfully implemented only in the longer run.

There should be more stringent requirements and greater scrutiny in granting and renewing concessions for plantations and for managing peatland use.

New technologies are often the solution to most problems, including the haze. For instance, water-bombing aircraft and land-clearing machinery would be made more efficient in years to come.

Lessons from economics would suggest the use of incentives and disincentives.

Since it is almost impossible to accurately determine which plantations started the fires, Indonesia should implement the law of strict liability. Instead of allocating blame to fire-starters, plantations should be held liable for any fire (above a certain scale) occurring on their land, regardless who started it.

This will create strong incentives for fire prevention and disincentives for clearing land with fire. Errant plantations should be fined or made to forfeit a liability deposit that they previously paid.

The magnitude of this penalty has to be carefully set as a determent, and fires must be accurately pinpointed with remote sensing and satellite technologies.

Land-owners should also be given grants, loans, training and logistical assistance to prevent or fight fires. If land-owners are unable to fight the fires themselves, they must be able to notify the local government for help before the fire gets out of control.

Gifts in kind — in the form of better roads, infrastructure, food vouchers and subsidies, or special funding for villages — could also be provided as incentives, if it can be shown that no fires above a prescribed size have occurred in a given area over a period of time.

Costs for the alternative methods of land clearing ought to be made known, so that Indonesia can negotiate and seek support from Asean and other countries in providing subsidies for firms and farmers who abstain from slash-and-burn practices. Alternative cost-efficient land-clearing practices could also be developed with increased investments in research and development.

Finally, this haze problem stemming from forest fires requires the involvement of multiple stakeholders, which include both culprits and victims. And since the daily greenhouse gas emissions from Indonesia's raging fires have reportedly exceeded the US' daily emission levels from all sources, the whole world is a collectively affected victim.

Indonesia could learn from transboundary pollution experiences in other regions, and seek support and guidance from international agencies such as the United Nations, World Health Organisation, World Wildlife Foundation and the Roundtable for Sustainable Palm Oil. The international community should also be concerned about the Indonesian fires, as this has obvious implications for global warming.

This will take time

We do not doubt that the haze problem will eventually be resolved, but it will take longer than three years for a discernable policy impact, especially when weather also plays a part.

Besides, Indonesian laws and the system of governance are complex and will require time to change. Resources in terms of manpower and funds have to be budgeted. As a large and spatially distributed country — 34 provinces subdivided into regencies, cities, districts and administrative villages, each with its own set of laws and regulations — we should not expect results soon, even with sustained political will.

Given that there are many suggested solutions, ultimately, the test is whether there will still be fires of such magnitude in the future. In the meantime, just as in the case for Singapore, it is important for Indonesia to estimate the aggregate and sectorial damages of the haze and fires, to itself and to the other affected countries.

With this information, the country can then determine whether the returns from such pollutive industries are substantially greater than the damages suffered, and better plan the allocation of land concessions. If the damage costs more than the amount of economic benefits reaped, a rethink is required in the further growth of these industries; perhaps Indonesia could then consider diversifying its exports away from goods closely related to the fires.

In the interim, we can be optimistic about a solution, provided there is improved coordination among government institutions, weeding out of corruption, regional and international cooperation, technological advancements and, most importantly, a continued search for pragmatic solutions.

Prof Quah is president of the Economic Society of Singapore and head of economics at Nanyang Technological University, where Mr Tan is a PhD research student.

29 Facts and Figures

Euston Quah* and Tan Tsiat Siong[†]

*Professor and Head of Economics, Nanyang Technological University
[†]Researcher, Nanyang Technological University

The widespread impacts of forest fires and the consequential transboundary haze pollution have been studied on various fronts. Forest fires and haze have, over the years, resulted in serious health risks, environmental damage and economic ramifications on perpetrator countries (mainly Indonesia), victim countries (including Indonesia, Singapore, Malaysia, Brunei, Southern Thailand), and even affecting the wider world. It is imperative to identify these costs, and even more pertinent to put a dollar value — an objective measure — to them. This chapter highlights some efforts in quantifying the costs of the Southeast Asian haze and Indonesian forest fires.

The Southeast Asian haze dates back to the 1970s (Lee, 2015), but it was not until the severe episode in 1997 that attention really began to grow on this air quality disaster. Pioneering valuation studies on the 1997 haze started by formally uncovering the damage cost of haze in Singapore to be between 163.5 million US dollars and 286.2 million US dollars (Glover and Jessup, 1999; Quah, 1999; Narayanan and Quah; Quah, 2002). Quah (2002) not only examines Singapore's health and economic costs, but extends the analysis to indirect damages in the forms of loss in visibility and views, as well as loss in recreational activities. The total damages stands at 0.18 percent to 0.32 percent of Singapore's 1996 gross domestic product (GDP), with tourism as the hardest-hit sector. Table 1 presents the summary findings from Quah (1999, 2002).

The fire and haze crisis recurred intermittently over the years since 1997, and numerous research have also been conducted since. Fires and

Table 1. Summary of Total Damage Costs in Singapore due to the 1997 Haze

Impact of Haze Damages	Upper Bound Estimation (USD)	Lower Bound Estimation (USD)
Direct damages		
Direct cost of illness	1,535,668	1,186,900
Self-medication expenses	678,943	678,943
Loss in earnings/productivity	2,068,109	1,907,341
Preventive expenditures	234,909	3,524
Total health damage	4,517,629	3,776,708
Loss in tourism	210,499,067	136,577,290
Loss in local business	N.A.	N.A.
	214,966,696	140,353,998
Indirect damages		
Loss in visibility and views	71,137,941	23,057,133
Loss in recreation activities	94,170	94,170
Damages to biodiversity	N.A.	N.A.
	71,232,111	23,151,303
Total damage cost	**286,198,807**	**163,505,300**
Damage costs per person	95.39	54.50
Damage costs per household	369.90	211.31
% of 1996 GDP	0.32	0.18

Source: Quah (1999, 2002).

regional haze of significant scale returned most recently in 2015. With increased emphasis on sustainability, and the urgency to tackle climate change, the World Bank published a comprehensive report titled, "The Cost of Fire: An Economic Analysis of Indonesia's 2015 Fire Crisis" (World Bank, 2016) quantifying the cost to Indonesia. The study considered a thorough list of items: forgone crops production revenue when reclaiming burned land, losses to biodiversity and ecosystem services, cargo shipping interruptions, school closures and more. It was concluded that the 2.6 million hectares of Indonesian land burned between June and October 2015 resulted in a cost of at least 16.1 billion US dollars (1.9% of Indonesia's 2015 GDP). In contrast, palm oil production only constitutes

Table 2. Estimated Losses and Damages from Forest Fires and Haze, June–October 2015 (USD Millions)

	Total Losses and Damages (USD Millions)
Agriculture	4,839
Estate crops	3,112
Food crops	1,727
Environment	4,253
Biodiversity loss	287
Carbon emission	3,966
Forestry	3,931
Manufacturing & mining	610
Trade	1,333
Transportation	372
Tourism	399
Health	151
Education	39
Firefighting costs	197
Total in USD million	**16.124**

Source: World Bank (2016).

12 billion US dollars of value-add, suggesting that converting valuable forest assets into agriculture land by fire is not at all economically efficient. Key findings are presented in Table 2.

The measured costs of fires and haze remain far from exhaustive. For example, airborne pollutants are said to increase the incidences of dementia, but the exact long-term impacts on health are still not evident. As such, the long-term impacts of productivity, economic activities and socioeconomic outcomes also remain uncertain. Quantifying the impacts on the ecosystem is an even greater challenge, as fires destroy natural genetic variability, reduce the land's capacity to store carbon, wipe out organisms, and create irreversible damage (World Bank, 2016). This might explain the discrepancy between the World Bank's estimation of 16.1 billion US dollars, and Indonesian President Joko Widodo administration's estimation of 35 billion US dollars (Chan, 2015).

Ongoing studies will continue to build our understanding of the fires and haze, and provide insights for policy response by the region, by Indonesia, and even globally. For example, knowing which sectors suffer the most from the haze allows governments to provide targeted assistance. Cost sharing between the region to fight fires, or incentivise mechanical land clearing, also requires comparing the costs of fires and haze with the mitigation costs. To the large extent that the Indonesian forest fires release carbon and aggravate climate change, valuing the damages can also justify global involvement. For instance, the 1997 haze episode alone exceeded the annual carbon emissions of Europe (Chander and Quah, 2014).

Besides quantifying the cost of fire and haze, multifaceted investigations into the issue must still be undertaken. A case to illustrate would be a recent medical study done by the Singapore General Hospital linking poor air quality to increased risks of cardiac arrests outside hospitals (Ng, 2017). Researchers have even claimed that fine particulate matter (PM 2.5) increases risk-seeking behavior in the face of losses, causes people to exhibit weaker prosocial preferences, and by logical extension, cumulatively leads to a greater propensity to commit crimes (Chew, 2017). Such results might be less convincing than others, but in years to come, rigorous research and debate will eventually deepen our understanding and drive the resolution of the issue.

References

Chan, F. (2015, October 11). $47b? Indonesia counts cost of haze. *The Straits Times*. Retrieved http://www.straitstimes.com/asia/47b-indonesia-counts-costs-of-haze.

Chander, P. and Quah, E. (2014, June 13). Tackling haze with cost sharing. *The Straits Times*. Retrieved http://www.straitstimes.com/singapore/tackling-haze-with-cost-sharing.

Chew, S. H. (2017, December 13). The haze's effects on cognition. *The Straits Times*. Retrieved http://www.straitstimes.com/opinion/the-hazes-effects-on-cognition.

Glover, D. and Jessup, T. (1999). Indonesia's fires and haze: The cost of catastrophe. Singapore: Institute of Southeast Asian Studies.

Lee, M. K. (2015, October 2). Haze in Singapore: A problem dating back 40 years. *The Straits Times*. Retrieved http://www.straitstimes.com/singapore/environment/haze-in-singapore-a-problem-dating-back-40-years.

Narayanan, S. and Quah, E. (Monograph). Economic damages from Indonesia fires and haze: Separating transfers from welfare changes.

Ng, A. W. Y. (2017, April 6). Haze brings risk of cardiac arrests: Study. *The Straits Times*. Retrieved http://www.straitstimes.com/singapore/health/haze-brings-risk-of-cardiac-arrests-study.

Quah, E. (1999). The economic and social cost of 1997 fires, in Lim, H. and D. Johnston (eds.), *Land-Forest Fires in Southeast Asia: Science and Policy*, World Scientific and National University of Singapore Press: Singapore.

Quah, E. (2002). Transboundary pollution in Southeast Asia: The Indonesian fires, *World Development*, 30(3): 429–441.

World Bank (2016). The cost of fire: An economic analysis of Indonesia's 2015 fire crisis. Jakarta, Indonesia: World Bank Group.

Printed in the United States
By Bookmasters